Actes du XIVème Congrès UISPP, Université de Liège, Belgique, 2-8 septembre 2001

Acts of the XIVth UISPP Congress, University of Liège, Belgium, 2-8 September 2001

ULg
UNIVERSITÉ de Liège

SECTION 3

PALÉOÉCOLOGIE / PALEOECOLOGY

I0085108

Sessions générales et posters

General Sessions and Posters

Édité par / Edited by

Le Secrétariat du Congrès

Présidents de la Section 3 :
Paul Haesaerts, Freddy Damblon

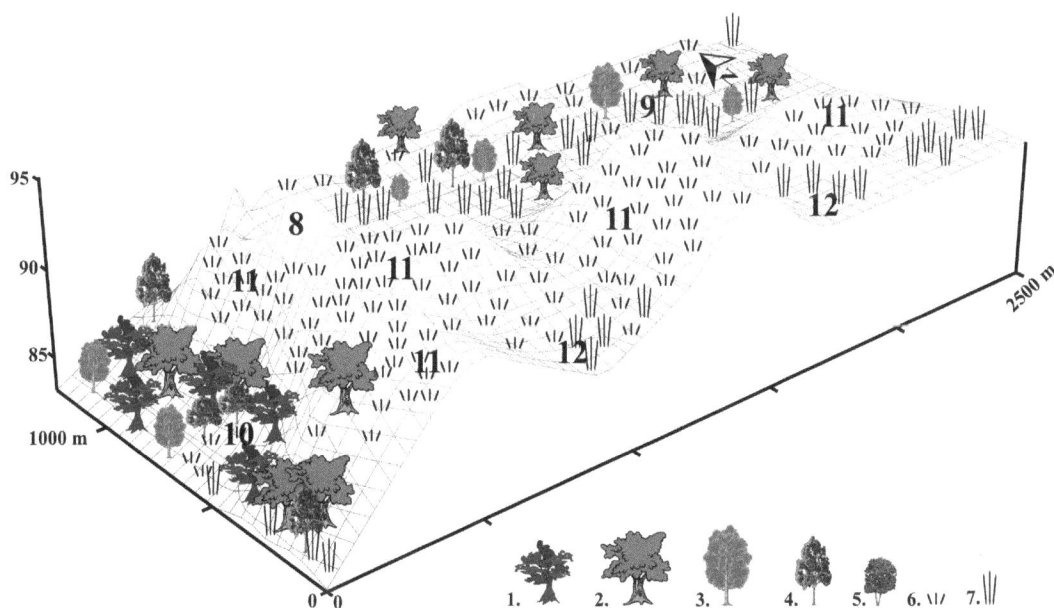

BAR International Series 1271
2004

Published in 2016 by
BAR Publishing, Oxford

BAR International Series 1271

Acts of the XIVth UISPP Congress, University of Liège, Belgium, 2-8 September 2001
Section 3: Paléoécologie / Paleoecology

ISBN 978 1 84171 624 4

Avec la collaboration du Ministère de la Région Wallonne. Direction générale de
l'Aménagement du territoire, du Logement et du Patrimoine. Subvention n°02/16341

Mise en page / Editing : Rebecca MILLER
Typesetting and layout: Darko Jerko

Marcel OTTE, Secrétaire général du XIVème Congrès de l'U.I.S.P.P.
Université de Liège, Service de Préhistoire
7, place du XX août, bât. A1, 4000 Liège Belgique
Tél. 0032/4/366.53.41   Fax 0032/4/366.55.51
Email : prehist@ulg.ac.be   Web : http://www.ulg.ac.be/prehist

BAR Publishing is the trading name of British Archaeological Reports (Oxford) Ltd.
British Archaeological Reports was first incorporated in 1974 to publish the BAR
Series, International and British. In 1992 Hadrian Books Ltd became part of the BAR
group. This volume was originally published by Archaeopress in conjunction with
British Archaeological Reports (Oxford) Ltd / Hadrian Books Ltd, the Series principal
publisher, in 2004. This present volume is published by BAR Publishing, 2016.

Printed in England

BAR
PUBLISHING

BAR titles are available from:

BAR Publishing
122 Banbury Rd, Oxford, OX2 7BP, UK
EMAIL   info@barpublishing.com
PHONE   +44 (0)1865 310431
FAX   +44 (0)1865 316916
www.barpublishing.com

TABLE DES MATIÈRES / TABLE OF CONTENTS

SECTION 3 POSTERS

MISE EN PLACE DES DÉPÔTS DE L'ABRI SOLUTRÉO-BADEGOULIEN DU CUZOUL DE VERS (LOT, FRANCE)

Stéphane KONIK & Bertrand KERVAZO

Résumé : L'étude géologique du remplissage de l'abri du Cuzoul de Vers met en évidence trois pôles dynamiques : action du gel, éboulisations gravitaires, apports depuis le Lotinondations. Trois phases peuvent alors être distinguées : une période d'action conjointe de sédimentation d'origine alluviale et de gélivation qui fragilise les parois ; une phase de rafraîchissement des calcaires qui a produit l'intercalation d'apports gravitaires tendant à supplanter le colmatage par les sables alluviaux les alluvionnements ; un épisode durant lequel gélivation et contributionapports de la rivière redeviennent simultanés. Seules les modalités de la fragmentation évoluent et l'on ne perçoit pas de changement des apports du régime de la rivière.

La conservation exceptionnelle de la séquence archéologique (30 niveaux entre Solutréen supérieur et Badegoulien) résulte de la conjonction de facteurs propres au site : morphologie de l'abri protégeant le remplissage des colluvionnements du versant, vitesse de sédimentation exceptionnelle autorisée par la nature des calcaires (bréchiques, diaclasés, gélifs ...), colmatage sableux uniformément réparti par alluvionnements calmes, processus perturbants peu actifs en raison de la faiblesse des pentes et de la texture très sableuse.

Abstract: Sedimentation of Cuzoul de Vers rock shelter – Solutrean and Badegoulian - (Lot, France).

Three dynamic poles have been underlined by the geological study of Cuzoul de Vers deposits : frost action, rock falls and river contribution. Three phases could be distinguished: first a period of action which is to be connected with sandy sedimentation from the river and frost action which made walls more fragile ; then a phase in which limestones were trimmed and thus brought about the insertion of screes which finally superseded the sandy plugging from the river ; eventually an episode in which frost action became simultaneous with the deposit of sands from the river. Only the fragmentation proceeding evolved so that no change in the river contribution could be perceived.

The perfect state of preservation of the archaeological sequence (30 layers between Superior Solutrean and Badegoulian) has come from the conjunction on the site of such determining factors as the morphology of the rock shelter (which protected the sequence of slope deposits), the exceptional sedimentation rate (bestowed by the very nature of limestone : conglomerate, fractured and frost clastic), and a sandy plugging evenly distributed, so perturbation processes that remained however little active owing to gentle slopes and to a very sandy texture of the deposits.

PRÉSENTATION

Le site du Cuzoul de Vers se trouve dans le Quercy, sur la bordure orientale du bassin d'Aquitaine (France). Il est situé dans la vallée du Lot, une quinzaine de kilomètres à l'est de Cahors, dans la partie concave d'un méandre encaissé.

L'abri, de dimensions modestes (10 m de long pour 1,50 m de profondeur), apparaît 8,50 m au-dessus de l'étiage actuel. Il a été creusé à la croisée de diaclases, dans un escarpement de calcaires bréchiques de l'Oxfordien - Kimméridgien exposé au sud.

Le gisement a été découvert à l'occasion du réaménagement de la route Cahors – Figeac et a fait l'objet d'une fouille de sauvetage dirigée par J. Clottes et J.-P. Giraud de 1982 à 1986. Fouillé sur 2,50 m d'épaisseur, il a livré une stratigraphie exceptionnelle avec 2 niveaux de Solutréen et 28 niveaux de Badegoulien. Plusieurs dizaines de structures foyères bien préservées ont été mises au jour ainsi que de multiples passées ocrées formant parfois des niveaux fins et continus.

PRINCIPALES DYNAMIQUES SÉDIMENTAIRES

La séquence se caractérise par la superposition de couches sub-horizontales, continues, peu épaisses (1 à 10 cm), formées de castines plus ou moins riches en sables. Des couches repères, plus sableuses ou, au contraire, plus clastiques se différencient (fig. 1).

Les analyses montrent qu'en dépit des nombreuses formations susceptibles d'avoir participé au remplissage - calcaires des parois, alluvions des différentes terrasses, argiles de colmatage des diaclases, dépôts à composante éolienne des environs, etc. - l'alimentation naturelle des sédiments se réduit à deux origines : les alluvions récentes du Lot et les calcaires de l'abri. Ce nombre restreint de sources d'approvisionnement est une conséquence de la morphologie de la barre calcaire, avec éperons rocheux encadrant l'abri et replat le surmontant, qui a isolé la séquence des dépôts de pente voisins et des terrasses perchées situées en retrait.

Les alluvions récentes du Lot déposées sous l'abri lors de crues se caractérisent par des sables quartzeux et micacés.

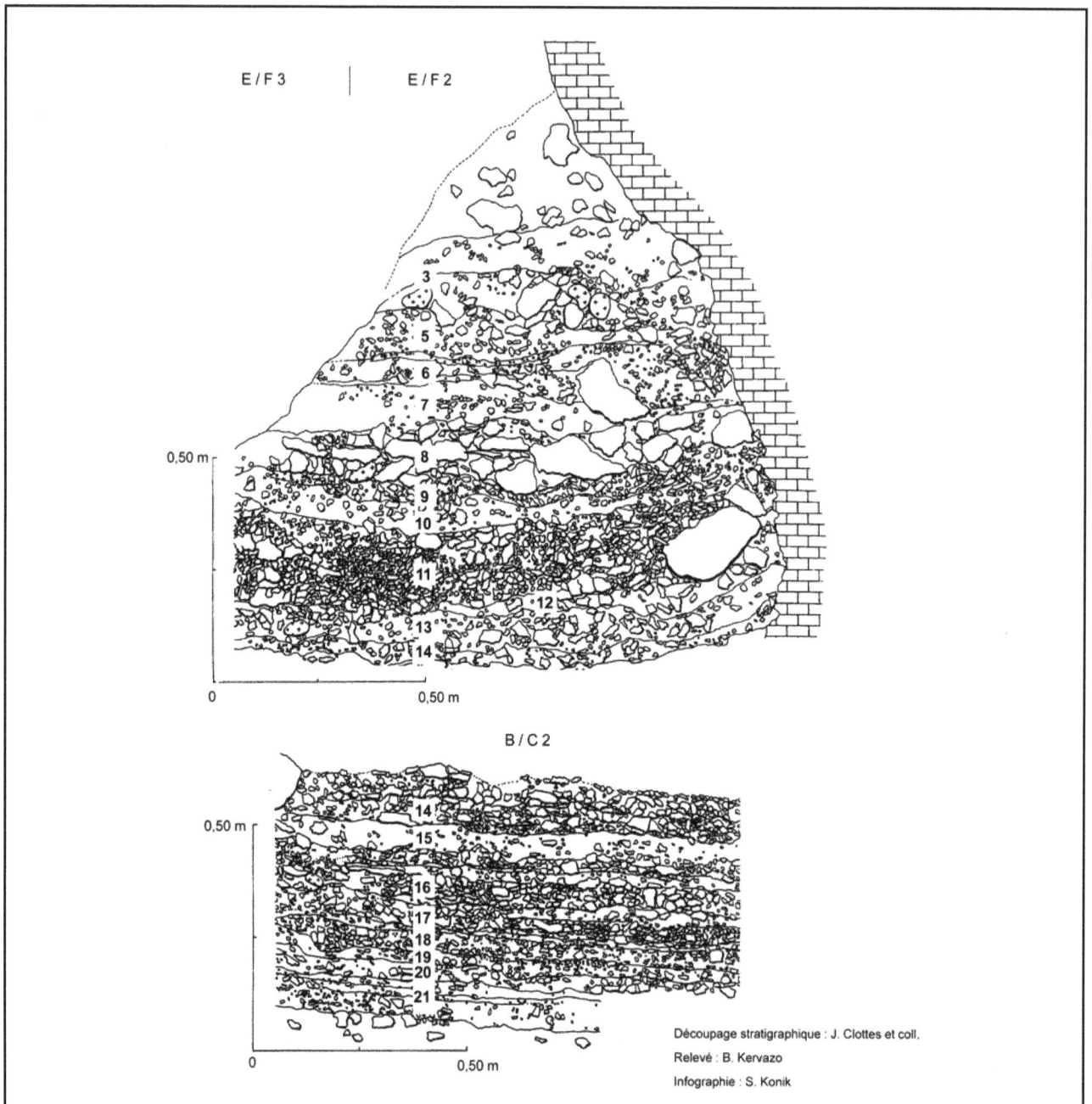

Figure 1 - Abri du Cuzoul de Vers : stratigraphie des couches supérieures et moyennes (coupes sagittales)

Elles sont à l'origine d'une part importante de la matrice. Elles ont pu être déposées directement sous l'abri lors de crues, ou indirectement, par reprise éolienne des dépôts de la plaine alluviale. Les couches qui leur sont rapportables, telle 15, sont plus ou moins fines, sub-horizontales et très étendues (fig. 1).

Les calcaires de l'abri sont à l'origine des clastes accumulés en formations plus ou moins ouvertes.

Deux processus de démantèlement se distinguent :

• Le cryoclastisme a largement contribué à la sédimentogenèse et les couches qui en résultent, telle 11, se retrouvent le plus souvent sur l'essentiel de la surface fouillée.

- Le cryoclastisme primaire a produit des clastes anguleux, homométriques, aux calibres moyens prédéfinis par le diaclasage et par la nature bréchique de la roche.

- Le cryoclastisme secondaire a généré des graviers, granules et sables calcaires, anguleux.

• La décompression est responsable d'accumulations localisées d'éléments hétérométriques, parfois volumineux, souvent émoussés sur une face comme dans la couche 8 par exemple (fig. 1). L'action du gel a pu se surimposer.

L'occupation humaine a sensiblement modifié les caractères de cette sédimentation naturelle : apports de vestiges, épandage d'ocre, aménagement de foyers, creusement de cuvettes, etc. Les analyses montrent que ces perturbations ont pu s'accompagner d'un enrichissement en limons, vermiculites et interstratifiés, qui peuvent trahir l'introduction sous l'abri, par l'homme, de constituants extérieurs.

ÉVOLUTION DE LA SÉQUENCE GÉOLOGIQUE

Aucune dérive sédimentaire ne se manifeste à l'intérieur de la séquence, comme le montrent les observations de terrain et les analyses. Ainsi, la granulométrie du sédiment traité par HCl ne révèle aucune évolution verticale des sédiments issus du Lot apports alluviaux : les courbes cumulatives demeurent remarquablement constantes, avec un mode compris entre 100 et 200 µm, ce qui conduit à un regroupement des échantillons sur l'image C-M (Passega, 1963) dans le secteur des suspensions uniformes, à l'exception de la couche 19, très anthropisée, et de la couche 11, très cryoclastique (fig. 2). De même, les courbes de distribution des carbonates en fonction de la granulométrie (Kervazo et Konik, à paraître) ne montrent aucune progression de l'impact du gel secondaire puisque les proportions des sables grossiers calcaires produits par microgélivation varient indépendamment de la succession stratigraphique (fig. 3).

En revanche, la juxtaposition des dynamiques - alluvionnements, cryoclastisme, décompression - est responsable de l'individualisation de trois ensembles (fig. 4).

Le premier, à la base de la séquence - couches 31 à 24 -, relativement homogène, se caractérise par des niveaux fins et réguliers. La permanence du colmatage sableux micacé associé à la microgélivation indique une régularité de l'alimentation à partir de par la rivière et des par les parois de l'abri.

Le deuxième correspond à la partie moyenne - couches 23 à 6 - et se caractérise par une superposition de dépôts ouverts et colmatés. Il reflète une sédimentation moins régulière qui a abouti à des couches mieux différenciées, moins planes, aux épaisseurs plus variables, pour lesquelles un processus de mise en place a prédominé : alluvionnement pour 15, gélivation secondaire pour 11, éboulisation gravitaire pour 8, par exemple. Les taux d'accrétions paraissent donc avoir été plus variables que précédemment, d'une couche à l'autre, voire à l'intérieur d'une même couche dans le cas d'éboulisations gravitaires localisées. Les dépôts ouverts montrent, de plus, que les apports des parois ont supplanté temporairement le colmatage par les sables fluviatiles et la microgélivation alluvionnement et par microgélivation.

Le troisième, enfin, correspond au sommet de la séquence étudiée - couches 5 à 1. Les dépôts redeviennent plus

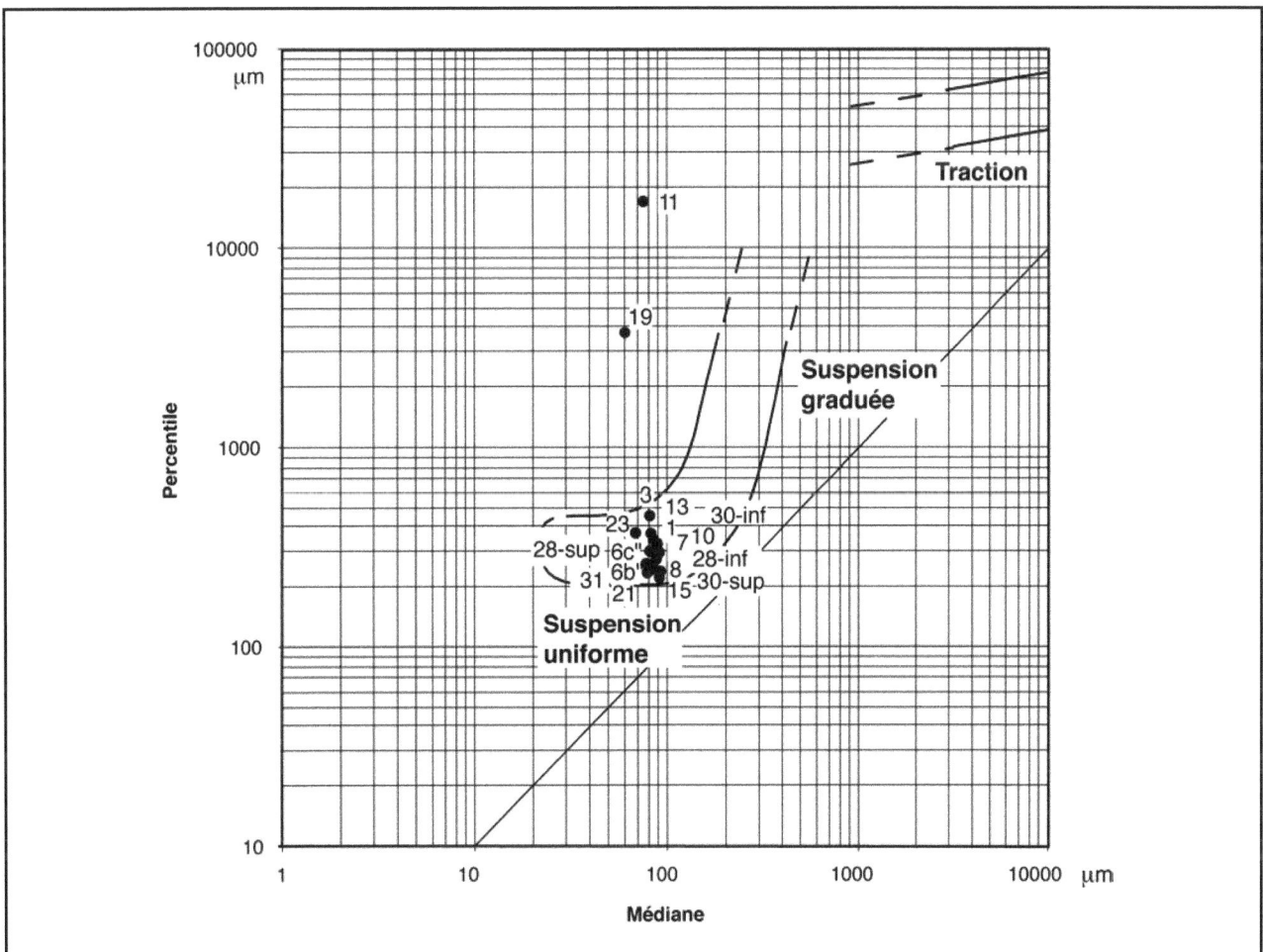

Figure 2 - Abri du Cuzoul de Vers : image C-M des dépôts

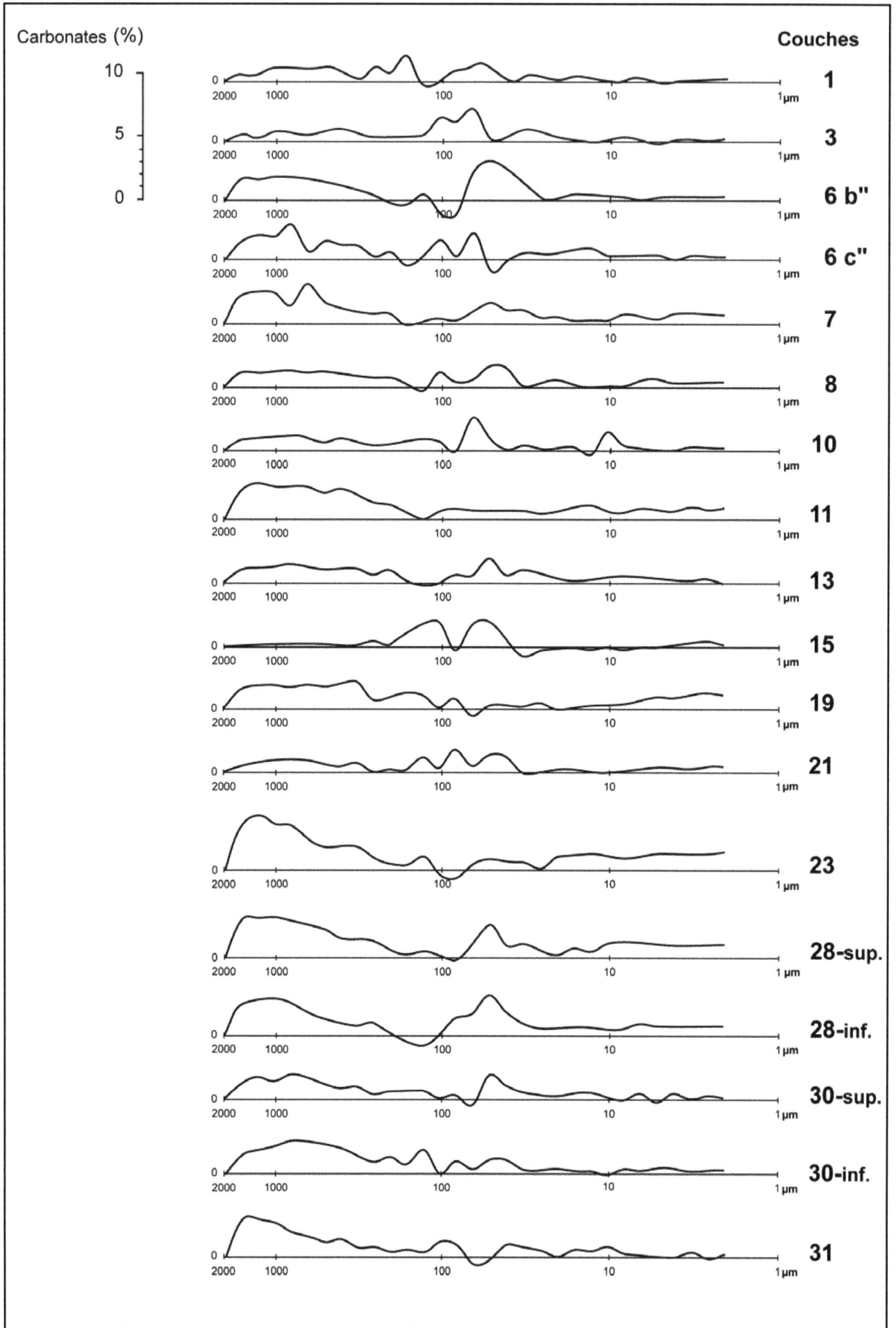

Figure 3 - Abri du Cuzoul de Vers : distribution des carbonates selon la granulométrie

homogènes et colmatés, rappelant l'alimentation régulière de la base, bien que des intercalations ponctuelles d'éboulis gravitaires, dues à des décompressions de parois, restent fréquentes.

Cette succession de trois ensembles évoque :

- d'abord une phase de fragilisation des calcaires par le gel, qui, d'après les âges [14]C (Clottes et Giraud, 1984 et 1985), a duré plus de 1000 ans (fig. 4) ;

- puis, conséquence de cette fragilisation, une phase de déstabilisation des parois par éboulisation gravitaire et cryoclastisme qui aurait duré près de 2500 ans ;

- enfin, un retour à la stabilité, renforcé par le fait que l'abri est en voie de régularisation, peu profond et, désormais, presque entièrement comblé.

L'évolution stratigraphique du remplissage semble donc avoir été essentiellement déterminée par les modalités du cryoclastisme et par la réaction des calcaires, indépendamment de la contribution alluviale directe (crues), ou indirecte (déflation de la plaine alluviale), qui présente des caractères sédimentaires remarquablement constants.

APPORTS ET LIMITES DE L'ÉTUDE GÉOLOGIQUE

L'étude géologique montre que le caractère exceptionnel de la séquence enregistrée (31 niveaux sur 2,50 m d'épaisseur) résulte d'une conjonction de facteurs favorables.

Ainsi, en ce qui concerne la sédimentation :

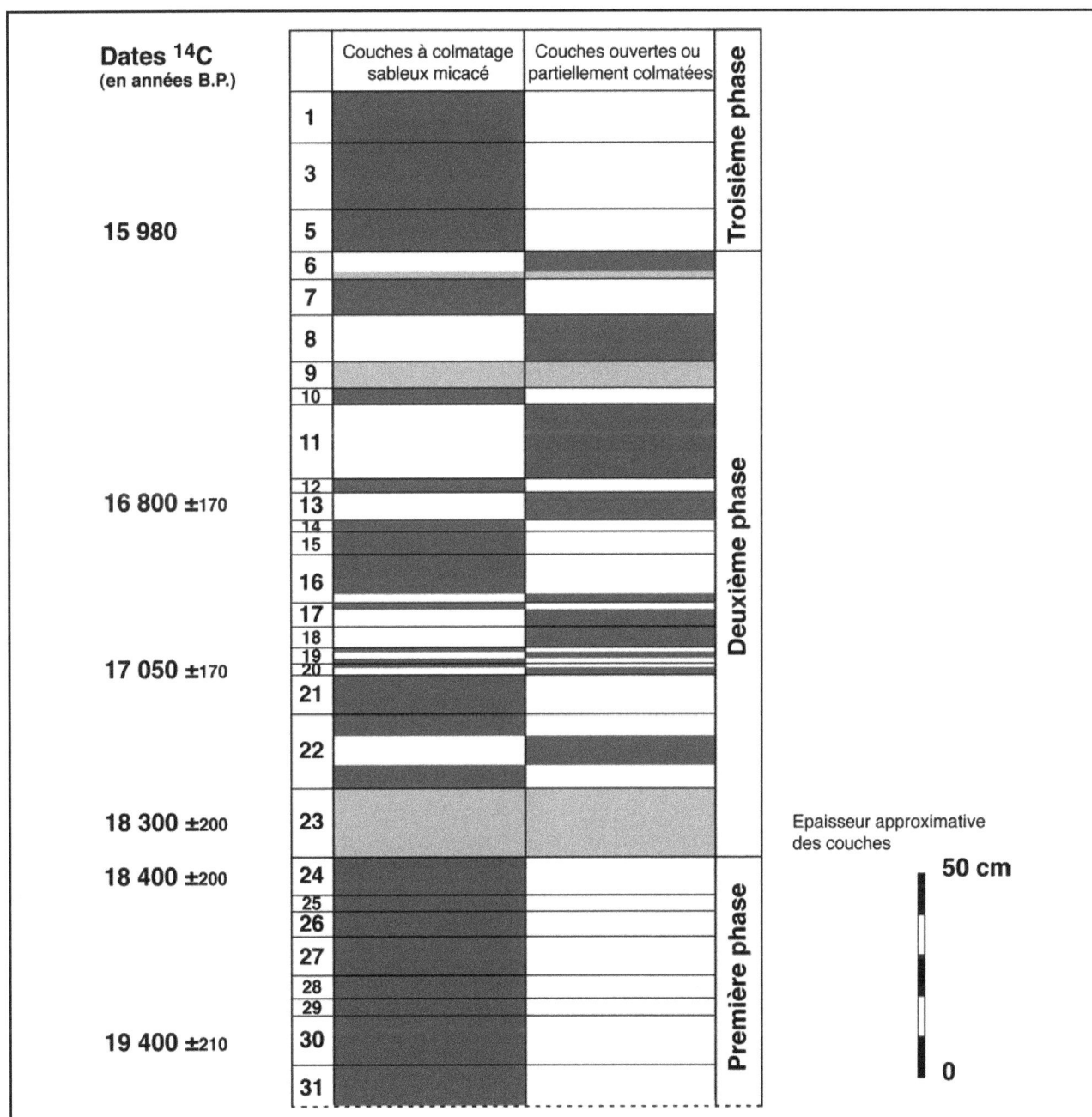

Figure 4 - Abri du Cuzoul de Vers : évolution de la sédimentation

- le cryoclastisme a été entretenu par les contrastes thermiques maximums, du fait de l'exposition au sud et par l'humidité occasionnée par les incursions de la rivière. Il a, par ailleurs, été facilité par la nature bréchique des calcaires, particulièrement propice au débitage en clastes aux calibres préétablis ;

- les éboulisations gravitaires ont été favorisées par le diaclasage du massif rocheux en contexte de versant escarpé et de surplomb d'abri ;

- la côte des crues a pu être accentuée par l'encaissement de la vallée et par la situation de l'abri en concavité de méandre, immédiatement en aval du point d'inflexion.

De même, en ce qui concerne la bonne conservation de sédiments aussi peu cohérents :

- les lames d'eau qui envahissaient le gisement 8,50 m au-dessus de l'étiage actuel et le vent n'ont pu, à chaque fois, sédimenter que des sables fins, transportés en suspension uniforme d'après l'image C-M. Sous l'abri, les crues elles n'ont pas occasionné d'érosion, alors qu'à l'avant, vers la vallée, les couches ont pu être recoupées en biseau ;

- le replat formé par l'avancée du seuil rocheux et l'horizontalité du colmatage sableux alluvial ont favorisé le maintien *in situ* du remplissage ;

- les éboulis gravitaires ont assuré la protection de certaines couches peu cohérentes, à l'image de 8, au-dessus de 9 et 10 par exemple ;

- enfin, le caractère parfois ouvert des sédiments et leur texture très sableuse ont été défavorables à d'éventuels remaniements liés au gel : aucune trace de cryoturbation, ni même de lentilles de glace, n'est apparue dans la séquence.

En revanche, l'étude géologique n'a pas permis de préciser plusieurs points importants.

Ainsi, l'action et les modalités du gel, notamment les fluctuations entre cryoclastisme primaire et secondaire, n'ont pu être interprétées, pour trois raisons au moins :

- ces fluctuations peuvent résulter de plusieurs paramètres : fréquence des cycles gel-dégel, profondeur du gel, vitesse d'enfouissement, humidité ambiante, etc. ;

- le diaclasage et la bréchification du calcaire ont brouillé d'éventuels signaux climatiques en facilitant la fragmentation et en imposant des calibres prédéfinis ;

- la pureté des sables et le caractère ouvert d'une partie des dépôts, impropre à la ségrégation de la glace, nous privent d'indices, par exemple sur la profondeur du gel dans le sol.

Malgré le rôle majeur de la fragmentation dans la sédimentogenèse, les variations observées ne revêtent donc, au mieux, qu'une portée stationnelle et aucun contrôle climatique global n'a pu être objectivement déduit de cette séquence, même à partir des trois phases identifiées. De telles limites sont habituelles dans ce type de remplissage et, plus généralement, dans les dépôts affiliés au karst (Campy, 1990, Ferrier et Kervazo, 1999).

Les variations du taux de sédimentation ne peuvent être appréhendées. Les éboulis ouverts, comme ceux de la couche 8, ont pu aussi bien résulter de brusques intercalations gravitaires dans une sédimentation d'origine alluviale constante, qu'au contraire révéler un ralentissement de cette dernière dû à un retrait prolongé de la rivière, puisqu'un niveau archéologique s'interstratifie. Remarquons que l'absence d'évolution verticale des sables quartzeux, alluvions qui plaide en faveur d'un rythme d'accrétion d'origine fluviatile relativement constant, contraste avec les taux d'accumulation habituellement plus fluctuants des processus gravitaires et cryoclastiques.

Enfin, le dénombrement des crues aux abords du dernier maximum glaciaire, un moment espéré en raison de l'identification ponctuelle de rythmes, n'a pas été possible. En effet, l'individualisation des épisodes s'est heurtée à la permanence de la composante sableuse qui peut aussi bien matérialiser des apports directs de la rivière, que des infiltrations dans les faciès ouverts, voire des d'éventuelles reprises des dépôts de la plaine alluviale par déflation en période de rhexistasie. L'intercalation de niveaux d'occupation, comme dans 15, montre, de surcroît, que la sédimentation des couches alluviales, même minces, a pu être complexe.

CONCLUSIONS

Cette séquence était, *a priori*, particulièrement favorable à une étude géologique : stratigraphie très détaillée et contrastée, dynamiques sédimentaires peu nombreuses et bien caractérisées, vestiges archéologiques et datations permettant un bon calage chronologique.

Malgré ces atouts, plusieurs questions fondamentales restent sans réponse. Citons la reconnaissance des variations de l'équilibre des processus, difficile à appréhender avec les méthodes actuelles, ou encore la durée nécessaire à la mise en place d'une unité sédimentaire, interrogation majeure pour la compréhension tant de la sédimentogenèse naturelle que des modalités de l'occupation humaine.

Les apports et les limites de l'interprétation des archives sédimentaires que constitue le remplissage d'un abri sous roche trouvent ici une bonne illustration.

Remerciements

J. Clottes, J.-P. Giraud, J. Jaubert et P. Chalard, sans qui ce travail n'aurait pu être réalisé ; B. Depelley pour la traduction du résumé en anglais ; N. Dauriac et A. Pignon pour la relecture et la mise en forme du manuscrit.

Adresses des auteurs

Stéphane KONIK
AFAN Grand-Sud-Ouest
Centre d'Activités «Les Echoppes»

156 avenue Jean Jaurès, Bâtiment F
33600 Pessac FRANCE
et
UMR 6042 du CNRS
Géodynamique des Milieux Naturels et Anthropisés, Maison
de la Recherche de l'Université B. Pascal
Laboratoire de Géographie physique
4 rue Ledru, 63057 Clermont-Ferrand. FRANCE

Bertrand KERVAZO
Ministère de la Culture et de la Communication
Centre National de Préhistoire
38, rue du 26ème R.I., 24000 Périgueux FRANCE
U.M.R. 5808 du C.N.R.S.

Bibliographie

BRAVARD J.-P., AMOROS C. et JACQUET C., 1986, Reconstitution de l'environnement des sites archéologiques fluviaux par une méthode interdisciplinaire associant la géomorphologie, la zoologie et l'écologie. *Revue d'Archéométrie*, 10, p. 43-55, 7 fig.

CAMPY M., 1990, L'enregistrement du temps et du climat dans les remplissages karstiques : l'apport de la sédimentologie. In *Remplissages karstiques et paléoclimats*, actes du colloque de Fribourg (Suisse), 13-14 octobre 1989. *Karstologia Mémoires*, 2, p. 11-22, 11 fig.

CLOTTES J. et GIRAUD J.-P., 1984, Rapport de sauvetage programmé : abri du Cuzoul à Vers (Lot). *Rapport interne S.R.A. Midi-Pyrénées*, 28 p., 8 fig., 7 ph., 1 tab.

CLOTTES J. et GIRAUD J.-P., 1985, Rapport de fouille de sauvetage programmé : abri du Cuzoul à Vers (Lot). *Rapport interne S.R.A. Midi-Pyrénées*, 24 p., 4 fig., 15 ph., 2 tab.

FERRIER C. et KERVAZO B., 1999, Réflexions sur la variabilité de l'enregistrement sédimentaire en entrée de grotte. In *Actes du Colloque européen - Karst 99*, Mende 10-14 septembre 1999, p. 89-94.

KERVAZO B. et KONIK S., à paraître, Etude géologique. *In* monographie consacrée au Cuzoul de Vers, *DAF*, env. 30 p., 8 fig., 8 tab.

PASSEGA R., 1963, Analyses granulométriques, outil géologique pratique. *Revue de l'Institut français du pétrole*, XVIII, 11, p. 1489-1499, 7 fig.

LA GROTTE DU PLACARD EN CHARENTE (FRANCE) : VARIABILITÉ DES FACIÈS SÉDIMENTAIRES ET CONSÉQUENCES SUR L'INTERPRÉTATION ENVIRONNEMENTALE

Bertrand KERVAZO & Catherine FERRIER

Résumé : Le remplissage de la grotte du Placard, étudié lors de fouilles récentes, illustre bien la variabilité des faciès et les différences dans l'enregistrement paléoenvironnemental qui peuvent exister en entrée de grotte.

Ainsi, les dépôts du secteur Galerie Louis Duport (GLD) se caractérisent par des clastes sédimentaires emballés dans des sables. Les calcaires sont très rares et les microstructures attestant le gel absentes. Ces caractères s'expliquent par le démantèlement d'anciennes alluvions karstiques piégées dans une diaclase dominant la séquence et par la situation reculée de ce secteur, responsable d'un amortissement des contrastes thermiques.

Dans la salle principale, au contraire, les calcaires sont extrêmement abondants. Le rôle prépondérant de la fragmentation résulte alors de l'extrême fissuration du massif rocheux et de la large ouverture du porche sur l'extérieur qui autorise la pénétration du gel.

Dans le secteur Y, en retrait, la fragmentation est plus réduite et le froid n'est enregistré que par une structure lamellaire locale.

Si les dépôts de la salle principale s'accordent apparemment avec les conditions du dernier maximum glaciaire, en revanche, dans GLD, aucun caractère du remplissage n'aurait permis une attribution à cet épisode, en l'absence d'industries.

Abstract: The sediments of the Grotte du Placard were studied in the course of recent excavations. They provide a good illustration of the variability and differences in the paleoenvironmental testimonies to be found in the entrances of caves.

Thus, the deposits in sector Galerie Louis Duport (GLD) are characterized by sedimentary clasts in the midst of sands. Limestone rocks are very rare and there are no microstructures testifying to frost. Those characteristics can be explained by the crumbling down of ancient karstic alluvions trapped within a diaclase overlooking the sequence and by the outlying place of that sector which causes the thermal contrasts to be deadened.

In the main chamber, on the other hand, limestone rocks are extremely abundant. Fragmentation is then mainly due to the importance fissuration of the massif and to the wide opening of its entrance which makes it possible to freeze inside.

In sector Y, in a recess, fragmentation is less active and cold episodes are only registered by a local superficial structure.

If the deposits in the main chamber apparently match the conditions during the last maximum glacial, on the other hand in GLD no characteristic whatever of the sediments would have allowed us to assign any layer to that episode, had it not been for their industries.

La grotte du Placard est mondialement connue pour la richesse de ses industries qui a notamment permis à l'abbé Breuil d'établir les subdivisions du Magdalénien ancien. A la fin des années 80, la découverte de gravures par L. Duport (Duport, 1990) a suscité la reprise de fouilles limitées sous la direction de J. Clottes (Clottes, 1994). L'étude géologique réalisée à cette occasion (Kervazo, Ferrier, O'yl, à paraître) a révélé le rôle déterminant des conditions stationnelles, à l'origine d'une grande variabilité de faciès. Ces variations ont des implications importantes sur l'interprétation paléoenvironnementale.

Le site se trouve une trentaine de kilomètres à l'est de la ville d'Angoulême, sur la commune de Vilhonneur. Il domine d'environ 10 m la Tardoire, petit affluent de la Charente. La cavité se développe dans une corniche de calcaires bajociens, micritiques à la base et oolithiques au sommet. Le plateau est recouvert par des placages d'argiles limoneuses rapportées à des colluvions sur la carte géologique (Le Pochat et coll., 1986).

La morphologie de la grotte a été guidée par la convergence de plusieurs diaclases. En effet, la cavité est constituée d'une Grande Salle dans laquelle aboutissent trois galeries (fig. 1) : le diverticule nord, le couloir René Laville et la Galerie Louis Duport (GLD). Nos travaux ont porté principalement sur le secteur GLD et sur son prolongement Y, qui ont conservé d'importants témoins sédimentaires. Quelques informations ont également pu être recueillies sur la Grande Salle, bien qu'elle ait été presque totalement vidée de son remplissage par les fouilles qui se sont succédé de 1876 à 1958. Ces trois secteurs renferment des séquences archéologiques contemporaines avec du Solutréen supérieur et du Magdalénien ancien.

VARIABILITÉ DES FACIÈS SÉDIMENTAIRES

Secteur GLD

La galerie GLD forme un conduit de 3,50 m de largeur pour 2 m de hauteur. Les dépôts archéologiques qu'elle renferme présentent un faciès très particulier, caractérisé par de nombreux clastes sédimentaires, parfois volumineux, emballés dans des sables micacés (photo 1 et 2). Ces éléments ont été approvisionnés par de vieilles alluvions karstiques,

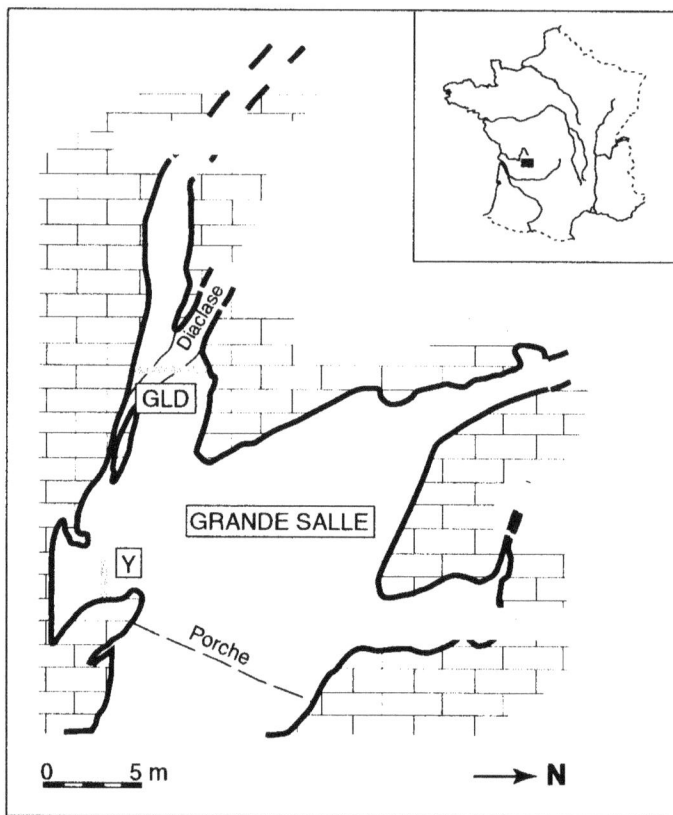

Figure 1. Situation du gisement du Placard et localisation des secteurs étudiés
(d'après J. Clottes et al.).

Photo 1 : secteur GLD. Clastes sédimentaires
emballés dans une matrice sableuse.

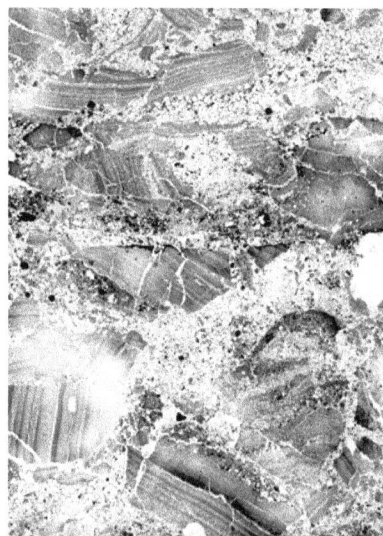

Photo 2 : secteur GLD. Micro-organisation en lame
mince des clastes sédimentaires (hauteur : 10 cm).

antérieures aux occupations humaines, encore visibles *in situ* à deux endroits de ce secteur :

– perchées dans une diaclase qui recoupe le plafond de la galerie (photo 3). Leur conservation en hauteur a été autorisée par une cimentation calcitique de leur base. Ces alluvions, ici de teinte orangée, sont constituées par l'alternance rythmique de lits sableux, micacés, à stratifications souvent obliques, et de lits limono-argileux, microlités (photo 4).

– sur une banquette qui forme la paroi sud de la galerie. Elles sont alors plus grises et plus massives.

La séquence étudiée a pu être subdivisée en trois ensembles sédimentaires (fig. 2) sur la base, notamment, de l'organisation des dépôts, de leur texture et de leur couleur (Kervazo, Ferrier, O'yl, à paraître) :

– Les ensembles inférieur et supérieur se caractérisent par des dépôts en berceau, des litages plus ou moins affirmés,

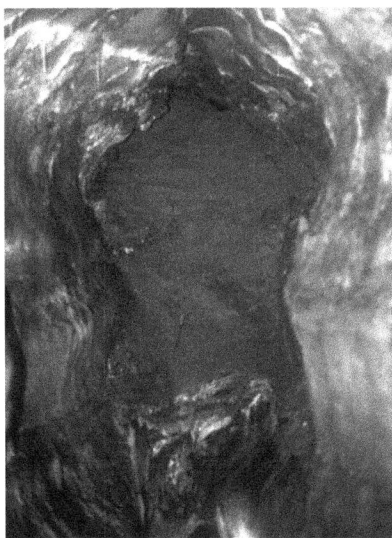

Photo 3 : secteur GLD. Alluvions karstiques
perchées dans la diaclase.

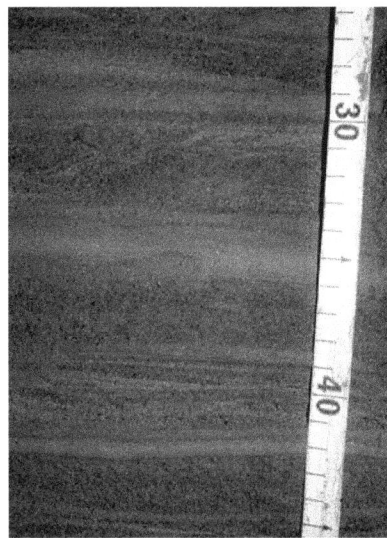

Photo 4 : secteur GLD. Alluvions perchées :
alternance rythmique de lits sableux et limon-argileux.

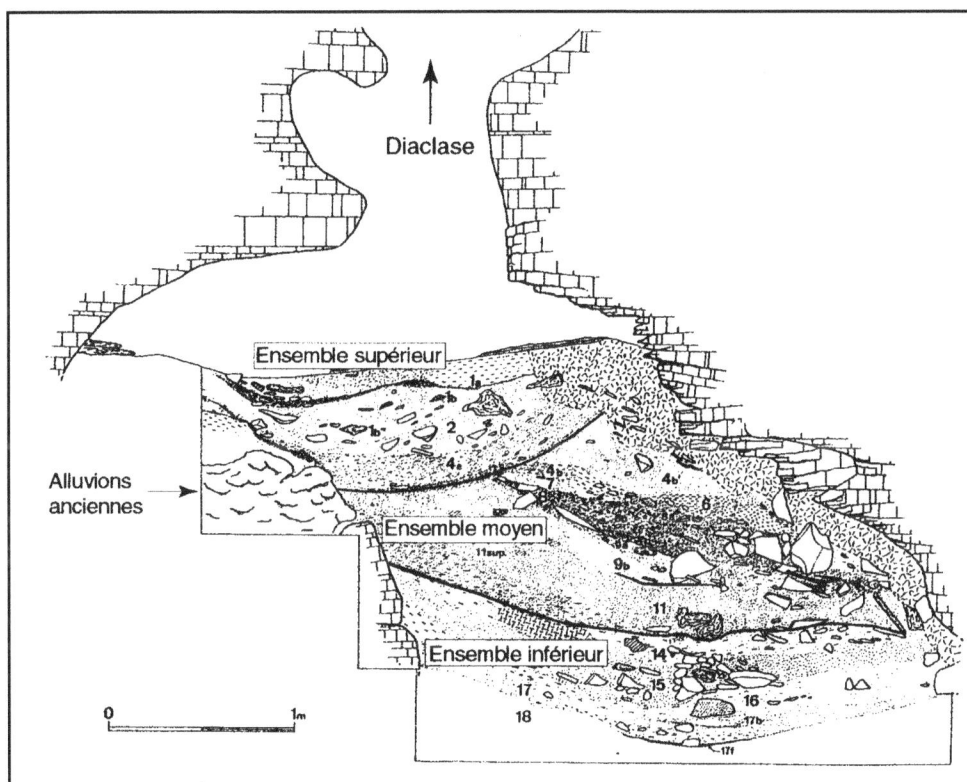

Figure 2. Secteur GLD. Relevé stratigraphique (d'après J. Clottes et S. Lacombe).

des clastes sédimentaires bien classés et de calibre millimétrique. Leur mise en place peut donc être rapportée, pour l'essentiel, à des ruissellements résultant d'écoulements longitudinaux dans l'axe du conduit. Ils ont remanié, à la base, des alluvions karstiques grises analogues à celles de la banquette et, au sommet, des produits de démantèlement des alluvions orangées de la diaclase.

– L'ensemble moyen s'individualise par une plus faible cohésion des sédiments et surtout par une organisation en demi-cône dont la pente atteint 30°. Il présente d'abondants clastes microlités orangés, hétérométriques, en position

désordonnée, emballés dans une matrice sableuse. Ces éléments se rattachent, par leur texture et leur couleur, aux alluvions karstiques perchées dans la diaclase : les clastes rappellent les lits limono-argileux tandis que la matrice est analogue aux lits sableux. L'origine du matériel et son organisation révèlent donc une mise en place régie par les éboulisations gravitaires à partir des formations meubles qui dominent la séquence.

Ainsi la majeure partie de la sédimentation de ce secteur repose sur deux processus : les ruissellements et les éboulisations gravitaires. En revanche, le démantèlement des parois est resté très limité. Il n'est guère perceptible qu'au nord, où les bancs du calcaire se délitent en minces plaquettes.

Secteur Y

Il correspond à un renfoncement de la paroi sud, dans le prolongement de GLD, à proximité du porche (fig. 1). Il est dominé par une cheminée, en partie colmatée par les vieilles alluvions karstiques, qui remonte vers la surface.

Le faciès des dépôts se caractérise de nouveau par de très nombreux clastes sédimentaires, auxquels s'ajoutent ici des cailloux calcaires assez abondants, parfois oolitiques, et des éléments provenant de l'extérieur, tels que des galets de quartz, des pisolithes ferrugineux ou encore de la kaolinite. Le microfaciès présente localement une fine structure lamellaire qui témoigne de l'intervention du gel.

Ici encore, les processus gravitaires sont responsables de l'essentiel de l'accumulation. Comme dans GLD, ils ont remanié les alluvions anciennes, mais dans ce secteur, ils ont également affecté la paroi et le plafond ainsi qu'en témoignent les fragments de calcaire oolitique. En outre, la proximité de la surface a autorisé l'infiltration de formations du versant par les fissures.

Grande Salle

Ouverte sur l'extérieur par un large porche, elle se caractérise par des parois hautes et presque verticales ainsi que par un plafond sub-horizontal conditionné par la stratification du massif rocheux. Malgré les bouleversements dus aux fouilles anciennes, plusieurs indices témoignent de l'abondance des blocs et des cailloux calcaires dans le remplissage :

– les descriptions des fouilleurs mentionnent de nombreux niveaux d'éboulis ;
– les déblais et les rares témoins bréchifiés qui subsistent sont extrêmement riches en fragments calcaires (photo 5).

Les processus de fragmentation ont joué de nouveau un rôle déterminant. Mais ici, en l'absence d'alluvions karstiques analogues à celles de GLD et Y, ils ont directement affecté les calcaires du plafond et des parois.

CONSÉQUENCES SUR L'INTÉRPRÊTATION ENVIRONNEMENTALE

Les trois secteurs étudiés, qui ne sont pourtant distants les uns des autres que de quelques mètres, présentent donc des faciès très différents, notamment conditionnés par la nature des roches mères disponibles et par la localisation des séquences. Cette diversité d'enregistrement, largement rapportable aux conditions stationnelles, a des conséquences sur l'interprétation paléoenvironnementale des dépôts.

Dans la Grande Salle, l'abondance des éboulis est due, en premier lieu, à l'intense fissuration du calcaire et à la faible épaisseur des strates (photo 5), particulièrement propices à l'action de la détente et de la gravité. Les grandes dimensions

Photo 5 : Grande Salle. Fissuration intense du calcaire, faible épaisseur des bancs et abondance des cailloux dans les déblais (premier plan).

du porche, en permettant une pénétration quasi instantanée du froid, ont également favorisé la gélifraction. La forte proportion des éléments grossiers, qui s'accorde avec les conditions du dernier maximum glaciaire, résulte alors au moins autant de facteurs d'ordre dynamique que climatique.

Dans GLD, les faciès ont été conditionnés par les alluvions karstiques, notamment celles en position instable dans la diaclase (photo 4). Le démantèlement et la gélifraction des parois sont restés très limités. Ceci résulte d'une part de la présence de ce colmatage alluvial ancien et d'autre part d'un amortissement des contrastes thermiques lié aux dimensions réduites et à la position reculée de la galerie. Aucun caractère du remplissage n'évoque ici la rigueur des environnements contemporains : les processus identifiés (éboulisations gravitaires et ruissellements) sont quasiment dépourvus de signification climatique.

Le secteur Y, relativement proche de l'entrée et de la surface, présente un faciès mixte qui comporte à la fois des calcaires, des clastes sédimentaires provenant des alluvions et des matériaux de l'extérieur. Sa position intermédiaire explique qu'il a été soumis à un cryoclastisme plus important qu'en GLD mais plus réduit que dans la Grande Salle. De plus, elle a permis à la fraction fine d'enregistrer les effets du froid, puisqu'une structure lamellaire s'est développée.

CONCLUSION

Cette étude illustre, s'il en était besoin, le rôle des conditions stationnelles en porche de grotte. Dans le site du Placard, elles ont conditionné des faciès sédimentaires dont la diversité a des conséquences importantes sur notre perception des climats contemporains. Outre les différences déjà largement soulignées dans l'enregistrement inter-site (Campy, 1990, Ferrier et Kervazo, 1999, par exemple), cette variabilité à l'intérieur d'un même gisement incite à la prudence vis-à-vis de l'extrapolation à l'échelle régionale des données paléoenvironnementales tirées de ce type de remplissage.

Remerciements

J. Clottes pour la traduction du résumé en anglais ; A. Pignon pour la relecture et la mise en forme du manuscrit.

Adresses des auteurs

B. KERVAZO
Ministère de la Culture
Centre National de Préhistoire
38 rue du 26ème RI 24000 Périgueux FRANCE
UMR 5808 du CNRS.
Email : bertrand.kervazo@culture.gouv.fr

C. FERRIER
Université de Bordeaux I
Institut de Préhistoire et de Géologie du Quaternaire
UMR 5808 du CNRS
avenue des Facultés, 33405 Talence cedex FRANCE
Email : c.ferrier@iquat.u-bordeaux.fr

Bibliographie

CAMPY, M., 1990, L'enregistrement du temps et du climat dans les remplissages karstiques : l'apport de la sédimentologie. In Remplissages karstiques et paléoclimats; actes du colloque de Fribourg (Suisse), 13-14 octobre 1989, *Karstologia Mémoires*, n° 2, p. 11-22.

CLOTTES, J., 1994, L'art pariétal paléolithique en France : dernières découvertes. *Completum*, 5, p. 221-233.

DUPORT, L., 1990, La grotte du Placard, commune de Vilhonneur (Charente). Découverte d'une galerie ornée Juillet 1988. *Bulletins et Mémoires de la Société Archéologique et Historique de la Charente*, p. 183-227.

FERRIER, C. & KERVAZO, B., 1999, Réflexions sur la variabilité de l'enregistrement sédimentaire en entrée de grotte. Actes du colloque européen « Karst 99 », 10-15 septembre 1999, *Etudes de géographie physique*, supplément n°XXVIII, CAGEP, Université de Provence, p. 89-94.

KERVAZO, B., FERRIER & C., O'YL, W., à paraître, Le Placard, étude géologique. In Monographie consacrée au gisement. 38 p.

LE POCHAT, G., FLOC'H, J.-P., PLATEL, J.-P. & REGOING, M. 1986, *Carte géologique de la France à 1/50 000, feuille de Montbron*, n° 710, Éditions du B.R.G.M.

THE NUTRITION AND SUBSISTENCE OF THE INDIVIDUALS OF GROTTA CONTINENZA DURING THE TRANSITION FROM EPIPALEOLITHIC TO MESOLITHIC AND NEOLITHIC (PALEONUTRITION, FAUNAL ANALYSIS AND MICROWEAR STUDIES)

Francesca BERTOLDI, Emiliano CARNIERI, Francesco MALLEGNI,
Michelangelo BISCONTI & Fulvio BARTOLI

Abstract: we present here a brief study on a site located in the Abruzzo region of Italy, with occupation layers dating from Upper Palaeolithic to Neolithic and Bronze age. We took into consideration the evidence shown by human and animal remains such as paleonutritional data and microwear analysis of teeth and we tried to compare all these data to better understand the palaeoeconomical picture of this site.

Résumé : nous presentons ici un étude sur un site localisé in Abruzzo, une région de l'Italie, qui a eté occupè dès Paleolithique superieur au Néolithique. Notre travaille comprend des donnés anthropologiques et faunistiques qui sont compareés a fin de mieux comprendre le cadre paléoeconomique du site.

THE SITE AND THE HUMAN REMAINS

The cave site of Grotta Continenza is located in the Abruzzo region, on Monte Labrone at 710 metres asl., close to the town of Trasacco and facing the Fucino basin. After the discovery of few pottery fragments, excavations started in 1978 and are still in progress. The cave appeared at first as a rock shelter and in 1981 a large inner chamber and a tunnel were brought to light.

The stratigraphic sequence comprehends Roman and Bronze age levels, a very thick Early Neolithic phase with Impressed Pottery (layers 2-24), Mesolithic layers with Sauveterrian lithic industries (25-29), and Epigravettian layers (30-41). Starting from layers 34 the so called "Bertonian" lithic industry was discovered.

Datations were obtained from several samples and levels: here we can quote:

layer 7: 6170+/-75 BP	layer 27:9330+/-100 BP
layer 20: 6590+/-75 BP	layer 28: 9680+/-100 BP
layer 25: 9490+/-100 BP	layer 32: 10280+/-100 BP
layer 26: 9100+/-100 BP	layer 34: 10230+/-100 BP

Inside the cave burials dating to Neolithic age were found, of which one indicated a cremation ritual (two children and an adult), the earliest in Italy, while the others belonged to a MNI of 30 individuals from the cave and the rock shelter.

Dating to the Final Epigravettian several human remains were recovered, some of them buried in stone circles. From layer 28-29 we recovered a female and at least other two adults, from layer 30 at least three individuals (one probably male and one female), from layer 30 another female, from layer 31-32 a male burial in stone circles and other sparse bone fragments belonging to two adults, and finally from layer 33-35 the burial of a male adult lacking of the skull was brought to light. The paleopathological analysis of the human sample revealed cases of periostitis, osteoarthritis, scoliosis, severe eburnation and osteophytosis on a patella, Harris lines, high degree of dental wear (Bartoli et al., 2001, Grifoni Cremonesi, 1998).

FAUNAL ANALYSIS

The faunal studies by Wilkens (1991) showed two distinct economic phases for the large faunal assemblage of Grotta Continenza. The more ancient one has been revealed by the Upper Mesolithic-Paleolithic sample while the more recent one testifies of the Neolithic subsistence. Regarding the Early Upper Paleolithic assemblage, faunal remains show an intense hunting of big mammals (*Equus hydruntinus*, *Capra*...) while during the later periods of Upper Paleolithic the prey was represented also by small mammals such as *Vulpes*, *Lepus*, *Felys*, *Meles*. During the Mesolithic hunting of the above mentioned mammals continued and the faunal remains include also those belonging to *Ursus*, *Bos*, *Canis lupus*, *Capreolus capreolus*, *Capra ibex*, *Martes martes* as it has been indicated by more recent studies.

Bird hunting was present during Paleolithic and Mesolithic times: during the first period the number of species is limited to *Anas*, *Nyproca* and *Spatola*, while during the Mesolithic age the number of hunted species shows a marked increase (*Anas platyrhyncos*, *Aythya ferina*, *Fulica atra*, *Perdix perdix*, *Alectoris graeca*, *Coturnix coturnix*, *Buteo buteo*, *Turdus merula*, *Colomba livia*, *Aquila chrysetos*, *Otis tarda*).

A very interesting feature is the importance of the exploitation of fish resources that started during Paleolithic

RELATIVE ABUNDANCE OF SELECTED VERTEBRATE TAXA IN THE FAUNAL ASSEMBLAGE FROM THE GROTTA CONTINENZA			
TAXON	LATE PALAEOLITHIC	MESOLITHIC	NEOLITHIC
Carnivora	9.21	22.2	1.28
Rodentia	0.4	0.33	-
Lagomorpha	8.43	9.09	2.24
Artyodactyla	70.24	62.94	68.56
Perissodactyla	1.20	-	-
Small ruminants	10.44	5.05	6.97
Large ruminants	-	0.33	-
Fish fragments	~600[a]	9368	785

Fig. 1. Abundance of mammalian taxa in the faunal assemblages from the Grotta Continenza. Mammalian data from Wilkens (1989-90), simplified. a New fish data from layer 32 only.

RELATIVE ABUNDANCE OF DIFFERENT MAMMALIAN TROPHIC GROUPS IN THE FAUNAL ASSEMBLAGE FROM THE GROTTA CONTINENZA			
	LATE PALAEOLITHIC	MESOLITHIC	NEOLITHIC
Large herbivores	81.88	68.32	75.53
All herbivores	90.71	77.74	77.77
Carnivores	9.21	22.2	1.28

Fig. 2. Relative abundance of different mammalian trophic groups in the faunal assemblage from the Grotta Continenza. Data from Wilkens (1989-90).

and Mesolithic, as testified by the large number of fish remains recovered in layers 32-41, and played an important role also during the Neolithic age. The species is *Salmo trutta* (trout) and it was fished in winter and spring during the Mesolithic and in summer during the Neolithic period (the fishing season has been recognized from the X-ray assessment of vertebral growth rings). Mesolithic faunal remains comprise a large number of molluscs (*Helix ligata*, *Leucostigma candidescens* convertita). The Neolithic economy since 6500-7000 BC relied on five domestic species: *Canis familiaris*, *Sus scrofa*, *Capra ircus*, *Bos taurus*, *Ovis aries* (fig. 1 and 2).

MICROWEAR ANALYSIS OF TEETH

We analysed the dental microwear of some individuals from Epipaleolithic and Mesolithic layers.

The study was made on buccal surfaces of molars. Silicon rubber (Provil-L, Bayer, Leverkusen, Germany) and acrylic resin (Araldite LY554 with catalyser HY956, Ciba-Geigy, Basel, Switzerland) were employed to produce replicas of teeth surfaces for microscope observations. Microwear patterns have been observed using metallographic microscope at several magnifications in order to assess the kind of diet. Dental surfaces images have been recorded on a floppy disk and analysed through a semiautomatic image-reading program (Microwear 3.0) (Ungar, 1994).

We took into consideration several factors, some of which have been used for a preliminary comparison. These are: length of scratches and their standard deviation, width of scratches and their standard deviation, preferential direction of the traces, major axis and their standard deviation, indices of the ratios, vertical/total scratches, horizontal/total scratches.

All the specimens show a majority of vertical scratches compared with horizontal ones, even if in different percentages. The differences among the specimens could reflect the adoption of different food acquisition strategies and microwear patterns of all specimens seem to indicate a omnivorous diet (Lalueza Fox and Pérez – Pérez, 1993; Pérez-Pérez et al., 1999).

Further SEM studies of dental microwear and statistical comparison with other specimens could help to confirm our hypothesis.

PALEONUTRITIONAL STUDIES

A very interesting field of investigation in the paleoanthropological research is the trace element analysis by AAS of human bone in order to quantify the contents of Strontium and Zinc and to discriminate between a carbohydrate-based diet and a nutrition rich in animal proteins. Strontium is very high in vegetables, cereals, and certain kinds of fish, while Zinc content is very high in red meat and milk (Bisel, 1980; Klepinger, 1984; Schoeninger, 1981; Sillen e Kavanaugh, 1982; Turekian e Kulp, 1956; Underwood, 1977). Relating these two elements to Calcium and, in the case of Strontium, also to the quantity present in an herbivore from the same site we obtain a picture of the diet and nutrition followed by a human group in the past (Bisel, 1980). This paleonutritional picture can be usefully related to other archaeological evidence from the site such as faunal remains, artifacts, and to other results from the anthropological analysis such as paleopathological data (evidence of poor nutrition, dental pathologies) and microwear analysis of teeth. This is precisely what we tried to get for Grotta Continenza samples, that is paleonutritional analysis associated with faunal studies and microwear analysis of teeth. Our aim is to achieve an accurate insight into the diet and therefore into the subsistence and economy of a prehistoric human group. What is more important we tried to follow the changes in time from an Epipaleolithic-Mesolithic way of life (based on small animals hunting, fishing and gathering) to a Neolithic management of resources through agriculture and animal husbandry. Paleoenviron-mental data should not be forgotten such as the closeness of this site to the Fucino lake basin that could offer abundant freshwater fish and birds.

Finally our results have been compared with those obtained for other Paleolithic and Neolithic sites of Central and Southern Italy that have been analysed in our Laboratory (Bartoli, 1996; Bartoli et al., 2001).

The samples and results of our AAS analysis are presented in Table 1 and 2 and in Graph 1 and 2. The nine Final Epigravettian skeletons from Grotta Continenza have been compared to Gravettian and Epigravettian individuals and what is immediately evident from Graph 1 is that, while for the Gravettian comparison sample the values of Zn/Ca ratio generally go over that of Sr/Ca, thus indicating a diet relying on animal resources and a hunting-based economy, the Epigravettian samples from Romito start to show a higher percentage of Sr in the diet: a tendency that fully appears in the samples from Grotta Continenza. In this site the Sr/Ca ratio in human bones is higher than Zn/Ca in all the individuals except Continenza 9. Considering that Sr is extremely high in vegetables and cereals but also in certain kinds of fish and molluscs, we can therefore suppose that the economy of the site was relying on small mammals hunting but mostly on fish, molluscs and wild vegetal resources. As it has been confirmed also by faunal analysis (Wilkens, 1991), the "broad-spectrum revolution" had already started in Abruzzo around 10000 years BP. We have also to notice, regarding the adequacy of the nutrition that some individuals from the site show values of Sr and Zn that fall under the standard ones, thus maybe indicating seasonal exploitation of certain resources with alternating periods of adequate and poor nutrition. These data will be verified through further examination of skeletal stress indicators such as Harris lines or enamel hypoplasia.

Neolithic samples from the site compared to those from Southern Italy show a homogeneity of results, with Sr/Ca

Table 1. Upper Paleolithic samples: provenance, sex and age

Sample	Layer	Sex	Age
Continenza 1	T.28-29- Final Epigravettian	F	adult
Continenza 2	T. 28-29 - Final Epigravettian	M	adult
Continenza 3	T. 30 - Final Epigravettian	M?	adult
Continenza 4	T. 30 - Final Epigravettian	F?	adult
Continenza 5	T. 30 - Final Epigravettian	M?	adult
Continenza 6	T. 31-32 - Final Epigravettian	M	adult
Continenza 7	T. 31-32 - Final Epigravettian	F	adult
Continenza 8	T. 31-32 - Final Epigravettian	?	adult
Continenza 9	T. 33-35 - Final Epigravettian	M	adult
Paglicci 12	Gravettian	M	juvenile
Paglicci 25	Gravettian	F	adult
Parabita 1	Gravettian	M	adult
Parabita 2	Gravettian	F	adult
Romanelli	Gravettian	?	adult
Romito 1	Final Epigravettian	F	adult
Romito 2	Final Epigravettian	?	adult
Romito 3	Final Epigravettian	M	adult
Romito 4	Final Epigravettian	F	adult
Romito 5	Final Epigravettian	F	adult
Romito 6	Final Epigravettian	M	adult
Vado all'Arancio A	Final Epigravettian	M	adult

Table 2. Neolithic samples: provenance, sex and age

Sample	Sex	Age
Continenza 1	M	adult
Continenza 2	M	adult
Continenza 3	F	adult
Continenza 4	F	adult
Continenza 5	F	adult
Continenza 6	?	adult
Continenza 7	?	adult
Samari	(X males)	adult
Samari	(X females)	adult
Ripa Tetta	(X males)	adult
Ripa Tetta	F	adult
Monte Kronio	F	adult
Latronico	M	adult
Catignano	F	adult
Trasano	(X males)	adult
Trasano	(X females)	adult
Pulo di Molfetta	M	adult
Balsignano	M	adult

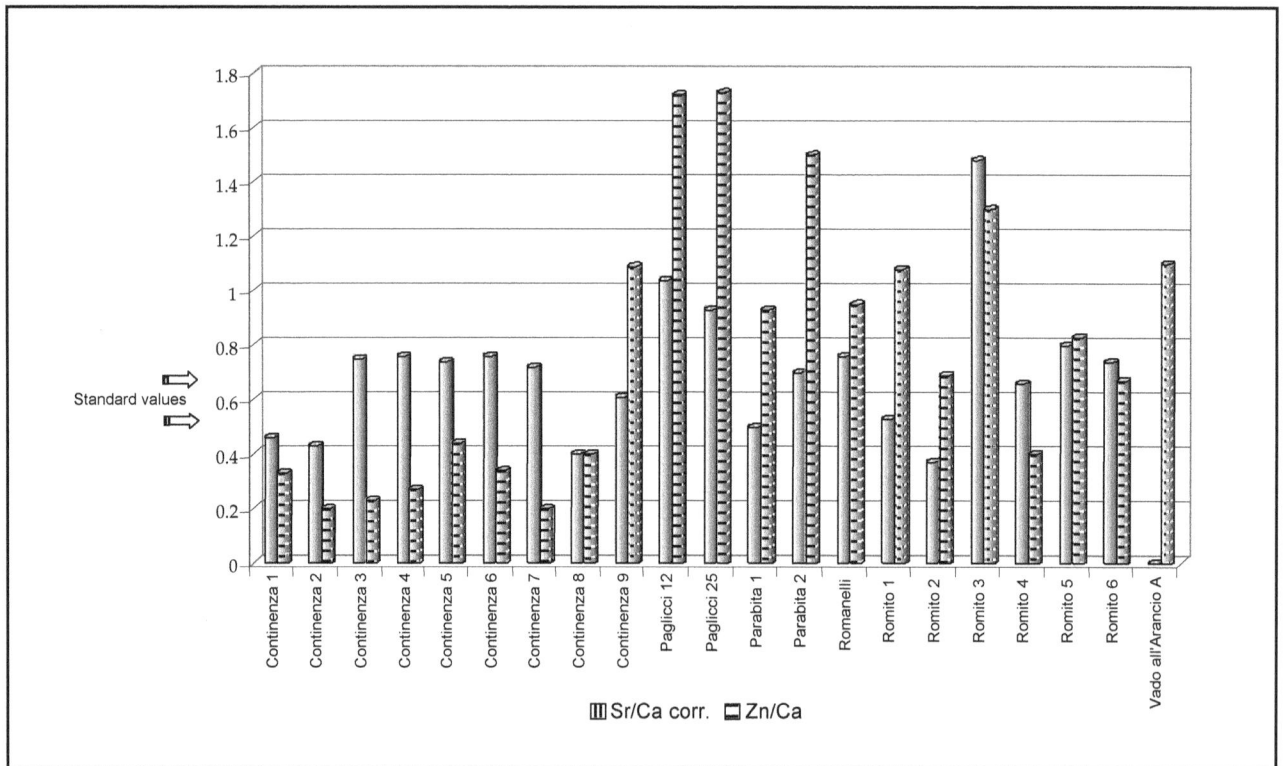

Graph 1. Values of Sr/Ca and Zn/Ca in Italian Middle and Late Upper Paleolithic

values higher than Zn/Ca, a result that is somehow expected given the importance of vegetal resources in agricultural economies. However we cannot forget, at least for our site, that fish exploitation resources played an important role also in Neolithic times.

Authors' address

University of Pisa
Dipartimento di Scienze Archeologiche
via S. Maria 53, 56100 Pisa ITALIA

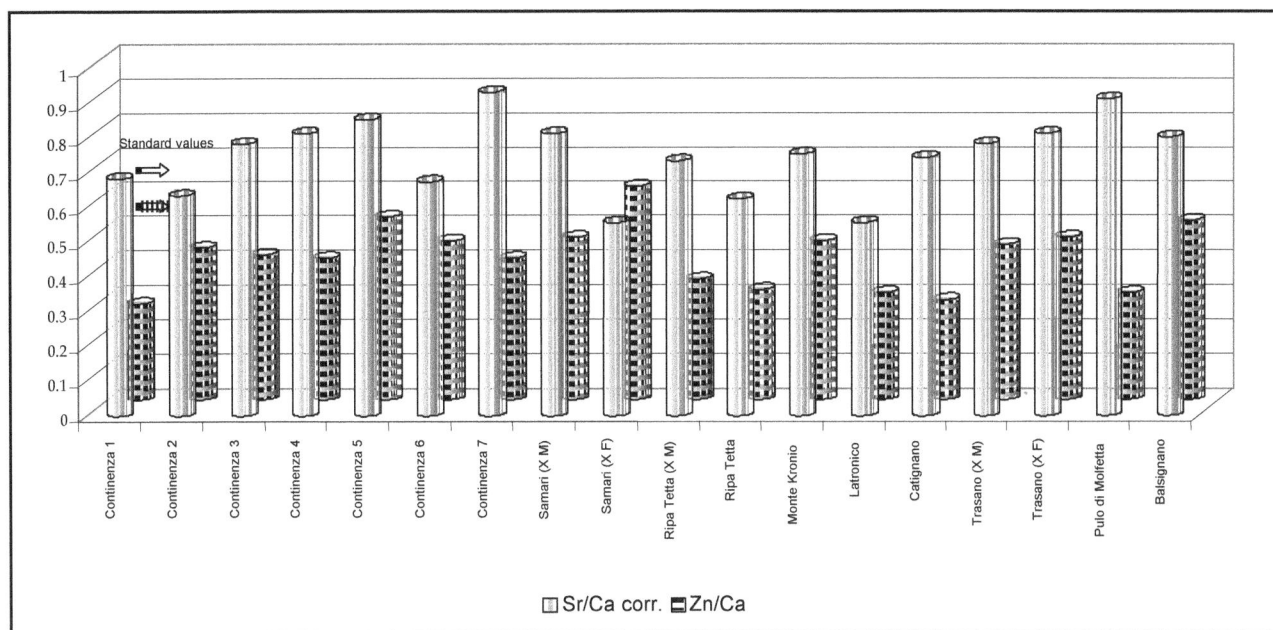

Graph 2. Values of Sr/Ca and Zn/Ca for Neolithic Italian Samples

Bibliography

BARTOLI F., 1996. Paleodieta. I gruppi umani neolitici dell'Italia centro meridionale. In: Forme e tempi della neolitizzazione in Italia meridionale e in Sicilia. Atti del Seminario internazionale di Rossano Calabro, Aprile-Maggio 1994, vol. 2: 499-507.

BARTOLI F., MALLEGNI F., BERTOLDI F., 2001. Analisi paleobiologica e paleonutrizionale dei resti scheletrici rinvenuti nella grotta Continenza. Atti del II Convegno "Il Fucino e le aree limitrofe nell'antichità", Celano, 1999.

BISEL S.C., 1980. A pilot study in aspects of human nutrition in the ancient eastern Mediterranean, with particular attention to trace mineral in several populations from different time periods. PhD dissertation, Washington.

CECCANTI B., 1994. Alterazioni diagenetiche dei reperti ossei nel terreno. In: F. Mallegni, M. Rubini, Recupero dei materiali scheletrici umani in archeologia, a cura di, Roma: 193-222.

FIDANZA F., LIGUORI G., 1988. Nutrizione umana. Idelson, Napoli.

GRIFONI CREMONESI R., 1998. Alcune osservazioni sul rituale funerario nel Paleolitico superiore della Grotta Continenza. Riv. Sc. Preist., 49: 395-410.

KLEPINGER L.L., 1984. Nutritional assessment from bone. Annual Review of Anthropology, 13: 73-96.

LALUEZA Fox C. and PÉREZ – PÉREZ A., 1993. The diet of the Neanderthal Child Gibraltar 2 (Devil's Tower) thrugh the study of vestibular striation pattern. J. Hum. Evol. 24: 29 - 41.

LAMBERT J.B., SZPUNAR C.B., BUIKSTRA J.E., 1979. Chemical analysis of excavated human bone from Middle and Late Woodland Sites. Archaeometry, 21: 115-129.

PÉREZ-PÉREZ A., BERMÚDEZ DE CASTRO J. M. and ARSUAGA J. L., 1999. Nonocclusal dental microwear analysis of 300,000 – years – old Homo heidelbergensis Teeth from Sima de los Huesos (Sierra de Atapuerca, Spain). Am. J. Phys. Anthrop. 108: 433-457.

RICHARDS M. P., PETTITT P. B., Stiner M. C., Trinkaus E., 2001. Stable isotope evidence for increasing dietary breadth in the European mid-Upper paleolithic.PNAS, 98, 11: 6528-6532.

SANDFORD M.K. 1992. A reconsideration of trace element analysis in prehistoric bone. In: Saunders S. e Katzenberg M., a cura di, Skeletal Biology of Past Peoples: research methods: 79-103.Wiley-Liss, New York.

SCHOENINGER M.J., 1981. Changes in human subsistence activities from the Middle East. PhD dissertation, University of Michigan.

SCHOENINGER M.J., 1981. The agricultural "revolution": its effects on human diet in prehistoric Iran and Israel. Paléorient, 7: 73-91.

SILLEN A., KAVANAGH M., 1982. Strontium and Paleodietary research: a review. Yearbook of Physical Anthropology, 25: 67-90.

TOOTS H., VOORHIES M.R., 1965 - Strontium in fossil bones and the reconstruction of food chains. Science, 149: 854-855.

TUREKIAN K.K., KULP J.L., 1956 - Strontium content of human bones. Science, 124: 405-407.

UNDERWOOD E.J., 1977 - Trace elements in human and animal nutrition. New York.

UNGAR P.S., 1994 A semiautomated image analysis procedure for the quantification of dental microwear II. Scanning 17: 57 – 59.

WILKENS B., 1991. Resti faunistici ed economia preistorica nel bacino del Fucino. Atti del I Convegno "Il Fucino e le aree limitrofe nell'antichità". Avezzano, 1989: 147-153.

PALEOECOLOGICAL DATA ON THE GRAVETTIAN SETTLEMENT
OF BILANCINO (FLORENCE, ITALY)

Biancamaria ARANGUREN, Gianna GIACHI, Marta MARIOTTI LIPPI, Miria MORI SECCI,
Stefano PACI, Anna REVEDIN & Giuliano RODOLFI

Résumé : L'attribution culturelle de l'habitat de Bilancino au Gravettien à burins de Noailles s'avère évidente sur la base des caractéristiques de l'industrie lithique, en grande partie représentée par des burins de Noailles de type classique, et de la chronologie absolue, autour de 25.000 ans b.p. La reconstruction du cadre paléo-environnemental s'appuie sur l'analyse des sédiments, des pollens et des charbones : L'association végétale révélée au travers des charbones et des pollens permet d'envisager la présence d'un paysage constitué de prairies avec une faible couverture arborescente et des zones marécageuses aux alentours avec un bois de conifères à proximité de l'habitat ; ceci pourrait indiquer un climat plus frais que le climat actuel; le spectre pollinique relève la présence notable d'herbes palustres.

La position de l'habitat dans une zone marécageuse, et la situation sédimentologique du niveau anthropique, englobé entre des niveaux de limons d'exondation permettent de supposer qu'il s'agit d'un campement saisonnier, car la zone ne devait être habitable qu'en période plus sèche.

Abstract: The Bilancino settlement is ascribed to the Gravettian culture (Noaillian facies) on the basis of typological characteristics of the lithic industry, mostly consisting of Noailles burins of classic shape, and of the absolute dating, about 25.000 bp. The reconstruction of palaeocological frame is depending on geomorphological, sedimentological, paleobotanical analysis (pollens and charcoals). The site was characterized by the closeness of water bodies, as indicated by the hydromorphic conditions in both the soil and alluvial-colluvial deposits, and by the presence of significant amount of hygrouphylous plants pollen grains. The forest cover was scarce. And was composed mainly of *Pinus sylvestris*, thus suggest that the climate would have been cold.

The scheme inferred induces us to think to a temporary man's presence, because during the rainy seasons the site was not suitable for settling.

THE BILANCINO SETTLEMENT (B.A. & A.R.)

The Bilancino settlement, in the Mugello Valley (NW of Florence, Italy), was discovered during digging works of the artificial lake of Bilancino; it was excavated in 1995-96 by the Archaeological Department of Tuscany (Aranguren & Revedin 1998).

The settlement is situated in a terrace of the Sieve river (238 metres above the sea level), near the confluence with the Stura torrent. Its extension was originally of about 300 square metres, but some parts were destroyed by natural events and by Bilancino dam works: two large areas (first and second sector) have been preserved. In the second sector a very large paleosurface (120 square metres) was discovered. The excavation revealed many flaking areas characterised by a certain variability of employed siliceous raw materials. A structured hearth was also found (Aranguren *et Alii,* 2001).

Among the three Bilancino dates by A.M.S. (Aranguren & Revedin, 1997), the most reliable is the oldest one, obtained analysing the charcoal samples of the structured hearth: Beta-106549 25410+ 150 BP.

The study of Bilancino lithic industry is still in progress: the analysis of the whole assemblage, composed of about 15.000 artefacts was just completed; among these the 10% are instruments (Laplace 1972 and 1964), the 1% are cores, 2% burin's spalls (Aranguren & Revedin 2001a).

The examen of the instruments shows that Bilancimo complex is chatarcterized by a high level of specialisation and of technological knowledge: the most represented typological group are Burins (70%), among them the largest part is constituted of Noailles Burins, with characteristics of classic type, as very small spalls and frequency of *encoches d'arrêt* (de Sonneville-Bordes & Perrot, 1956; Delporte 1968, DJINDJIAN F., 1977, DEMARS & LAURENT 1989). Few micro Gravette points, backed bladelettes and end scrapers are also present, while the substratum is pratically absent.

On the basis of the typological characteristic of the lithic industry, the settlement of Bilancino is to be ascribed to the Gravettian culture (Noaillian facies). The absolute age is also related to French industries of Perigordian Vc, dated about 25.000 B.P.

The reconstruction of palaeocological frame is depending on geomorphological, sedimentological, paleobotanical analysis; the sediment acidity has destroyed any faunal remains.

Samples for palinological and sedimentological studies were collected during the excavation in three different areas of the settlement (Cuts A, B, C) (Fig. 1). Carchoal samples come from the structured hearth in Q 36, that has been moved and then excavated in laboratory.

Studies of spatial analysis of artefacts and reassembling of chert flakes are still in progress, while a functional

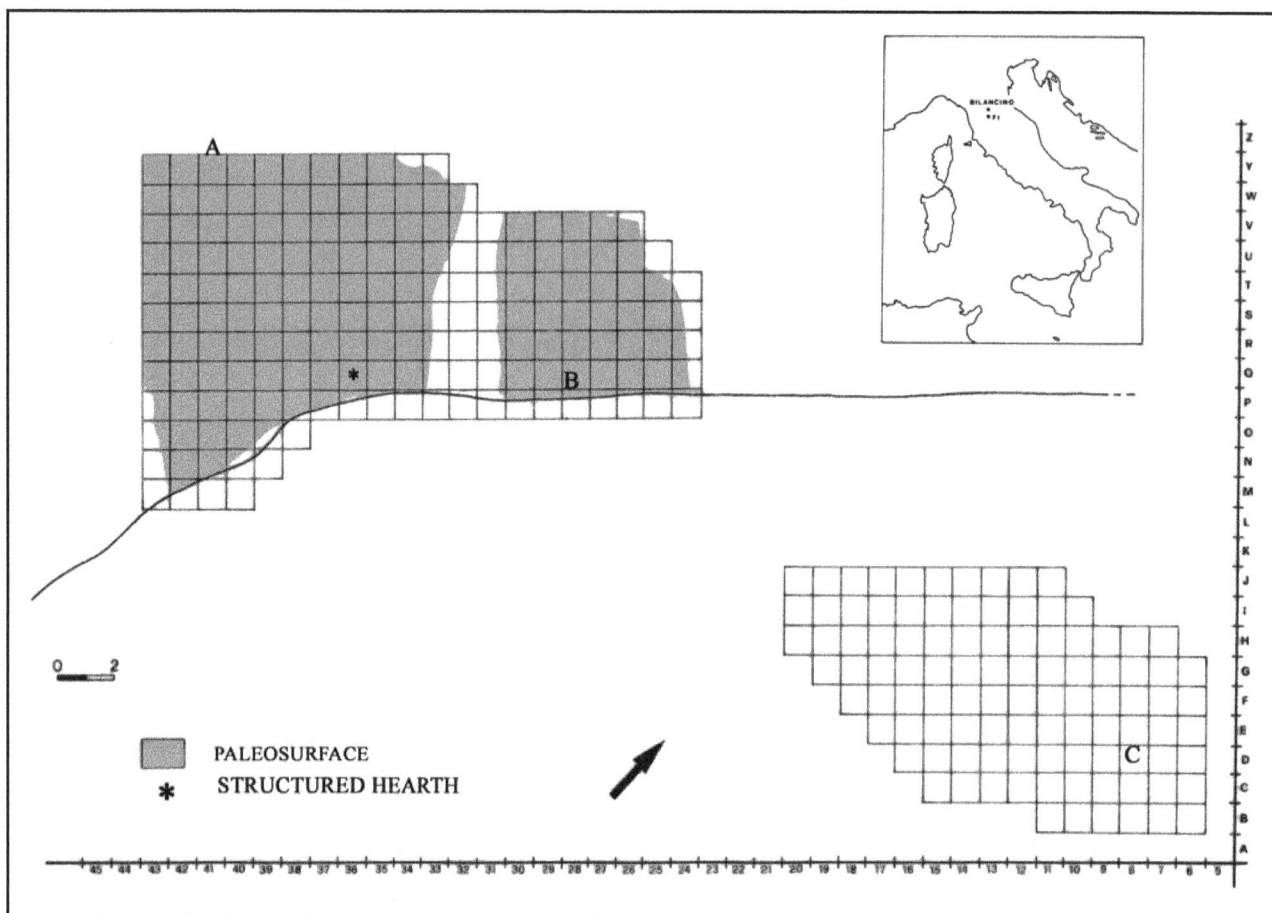

Fig. 1 - Bilancino: The excavation area with spotted samplings.

interpretation of the settlement was just hipothized: all available data, including archaeological documentation, ethnographical comparison, useweare analysis, in according with first paleoecological data, reconcile to a functional interpretation of this site as a seasonal camp for the harvesting and the processing of hygrophilous herbs. The functional interpretation of this settlement suggests a new vision on the scientific debate regarding the Noailles Burins lithic industries (Aranguren & Revedin, 2001b).

LOCATION AND GENERAL GEOMORPHOLOGICAL FEATURES OF THE SIEVE RIVER VALLEY BOTTOM UPSTREAM FROM BILANCINO GAP (G.R.)

The site is located at the northern margins of the wide River Sieve valley bottom, about 2 km upstream of the Bilancino gap. Since several years the whole valley bottom has been submerged by the waters of the same artificial lake, so that the present description refers to the former landscape, existing before the beginning of the works which have completely remodelled the morphology of this stretch of valley bottom before the definitive submersion.

The general features of the landscape generally follow the geological structure of this stretch of pre-Apennine: the Barberino basin, in facts, constitutes a minor tectonic

depression in comparison to that, wider and deeper, of Central Mugello, even if both of them were occupied continuously by the great Mugello Villafranchian lake. From the structural point of view, they remained, however, separated by a rocky diaphragm, formed by Cervarola sandstones (pelitic-arenaceous flysch, here dominated by pelites), nowadays represented by the Colle Barucci – Poggio Campiano – Poggio Mausoni ridge, later "cut" by Sieve River (Bilancino gap) and its tributary Tavaiano Torrent (Ghiereto gap). Therefore, the Bilancino rocky "treshold" has constituted for longtime, after the filling up of the Villafranchian lake, about 900,000 years ago (Sanesi, 1965) the local base level for the whole fluvial network upstream of it, allowing the formation of the above mentioned alluvial valley bottom.

This latter, however, was far from forming a unique, flat surface. One can observe, in facts, the presence of the two lowest alluvial surfaces which characterize, much more extensively, the valley bottom of the Sieve River in the Central Mugello. (Sanesi, 1965; Rodolfi et Alii, 1978; Rodolfi, 1982; Benvenuti, 1995): the one built during the last glacial period (Würm of the old alpine terminology) and the post-glacial (Holocene) one. Both surfaces occupy the left side of the Sieve River; they are separated by a low (about 2 m) scarp from C. Colombaiotto to C. Il Piano, and then juxtapose upvalley along the stream of the Stura Torrent. Along the Sieve R. right side, on the contrary, the footslope of reliefs is directly connected to the actual flood plain (Fig.2).

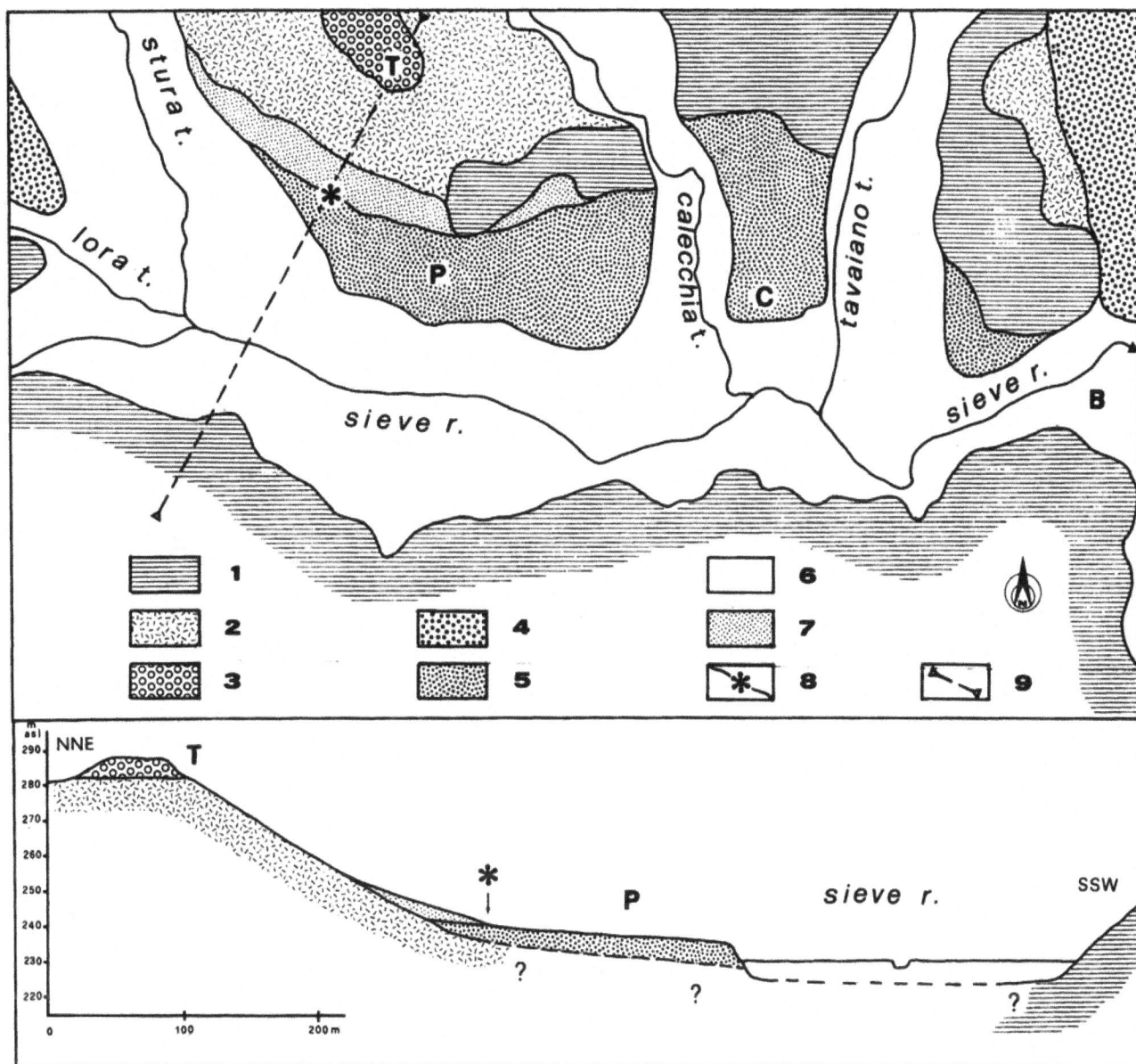

Fig. 2 – Geological sketch map of the Sieve R. flood plain and adjacent areas, upstream of Bilancino gap.
1 – Cervarola sandstones (Langhian-Oligocene). 2 – Lacustrine clayey and silty-clayey deposits (Villafranchian).
3 – Pre-penultimate glaciation (Mindel) alluvial deposits. 4 – Penultimate glaciation (Riss) alluvial deposits.
5 – Ultimate glaciation (Würm) alluvial deposits. 6 – Holocene (recent and actual) flood plain deposits.
7 – Recent and actual colluvial deposits. 8 – Location of the "Il Piano" archaeological site.
9 – Geological profile. B – Bilancino gap. C – Colombaiotto. P - Il Piano. T – Il Turlaccio

The remnant of a more ancient alluvial surface, elongated in a NW-SE, occupies the summit of a ridge separating the valley of the Stura T. from that of the Calecchia T,. between the localities of Moriano and Turlaccio, this latter directly hanging on the study site. It's a matter of a residual limb of the oldest alluvial surface of the Mugello basin, built during the Mindel glaciation (Sanesi, 1965); the related sediments, prevalently gravelly, could have been weathered during the successive interglacial (Mindel-Riss), originating a very leached tropical red soil. On the contrary, any track, on the study area, of another alluvial surface, that formed during Riss glaciation (Sanesi, 1965, Rodolfi *et Alii*, 1978) extensively present just upvalley (Villa Palagio, Cavallina) between the Sieve R. and Lora T. riverbeds.

On the slopes of the Moriano-Turlaccio ridge, below the Mindelian reddish alluvial deposits, the lacustrine Villafranchia sediments outcrop, constituted by alternating levels of sand, silt clay and intermediate fractions, with prevalence or the finest ones. This area is significatively indicated by the toponym "Grete del Lago" (Lake Clays). In correspondence of the locality Il Fangaccio (toponym tied to the presence of fine lacustrine sediments as well), the bedrock with Cervarola sandstones comes to outcrop again.

Right of Sieve R. the terraced surfaces are not present, and the slopes are directly connected, by a sharp slope break, to the actual flood plain.

Detailed geological conditions

The archaeological site (cut A) is located in correspondence of the morphological passage between the South-facing slope of the Turlaccio ridge and the so-called "Wurmian surface" above mentioned, of alluvial origin. Such passage is characterized by a concave profile, which denotes the presence of a continuous belt of colluvial materials, deposed during time by runoff processes affecting both the Turlaccio weathered gravels and the underlying fine lacustrine sediments.

The reference-profile is the following, from the bottom to the top:

I. greyish clay and silty clay of the Mugello Villafranchian lake;

II. weathered lower alluvial gravels, discordantly lying (by an erosion surface) on the above mentioned formation;

III. silt of probably alluvial origin (flooding silt) containing flints and charcoal (anthropic level of 25,000 years ago):

IV. weathered and mottled flooding silt (this flood episode buried the anthropological level);

V. upper alluvial gravels, which "close" the alluvial activity on the Wurmian surface;

VI. recent and actual colluvial deposit.

Moving towards East (Cut C), this picture comes to a change: in facts, the presence of colluvial deposits interfingered with alluvial layers has been surveyed. Such material coming from upslope have not certainly been continuous, but tied with the temporal distribution of rainfalls, and therefore with climatic variations, even of long period. Sediments deposed by Sieve R. alternated to them during frequent flood episodes; the presence, just downstream, of the Bilancino "threshold" made easy then, as well as till some years ago, these phenomena of overdeposition along the riverbed and continuous flood plain accretion.

In other words, in the study site alternated depositions of colluvial and alluvial material have taken place, at least up to the end of the last glaciation, and before the renewal of the vertical erosion in the riverbeds, which marked the beginning of Holocene.

The permanence of environmental conditions dominated by the presence of water is demonstrated by the hydromorphic conditions in both the soil and alluvial-colluvial deposits, characterized by frequent mottles, testifying the past presence of a fluctuating water table.

Cut A - Grain size analysis

sample nr / depth cm	gravel	coarse sand	fine sand	silt	clay
A5 130	4,0	8,0	25,0	32,0	31,0
A6 140	25,0	17,0	20,5	15,5	22,0
A7 150	0,0	5,5	32,0	22,5	40,0
A8 160	6,0	14,0	30,0	18,0	32,0
A9 170	45,0	13,0	18,0	11,0	13,0
A10 180	5,0	16,0	26,5	24,5	28,0
A11 190	0,2	6,3	23,7	37,3	32,5
A12 200	0,2	4,8	25,0	32,5	37,5

Cut C - Grain size analysis

sample nr / depth cm	gravel	coarse sand	fine sand	silt	clay
C1 135	0,0	3,0	22,0	31,0	44,0
C2 140	0,0	2,0	23,0	31,0	44,0
C3 145	0,0	3,0	24,0	32,0	41,0
C4 150	0,0	2,5	23,5	34,0	40,0
C5 155	0,0	4,0	21,5	30,0	44,5
C6 160	0,0	4,5	26,0	28,0	41,5
C7 165	0,0	4,5	30,5	28,5	36,5
C8 170	0,0	5,0	30,0	28,0	37,0
C9 175	0,0	3,5	29,5	32,0	35,0
C10 180	0,4	4,1	29,5	32,5	33,5
C11 185	0,3	3,7	30,0	32,0	34,0
C12 190	0,0	3,0	33,0	30,0	34,0

Cut A - grain size analysis

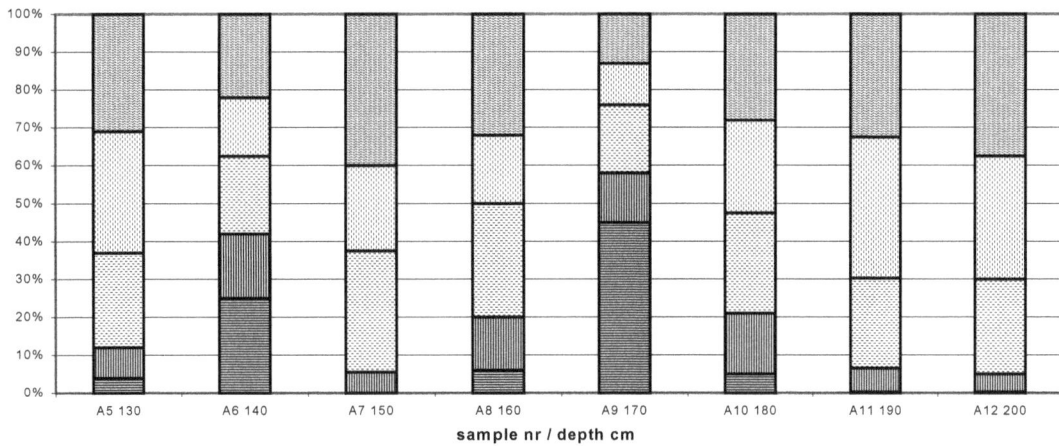

sample nr / depth cm

Cut C - grain size analysis

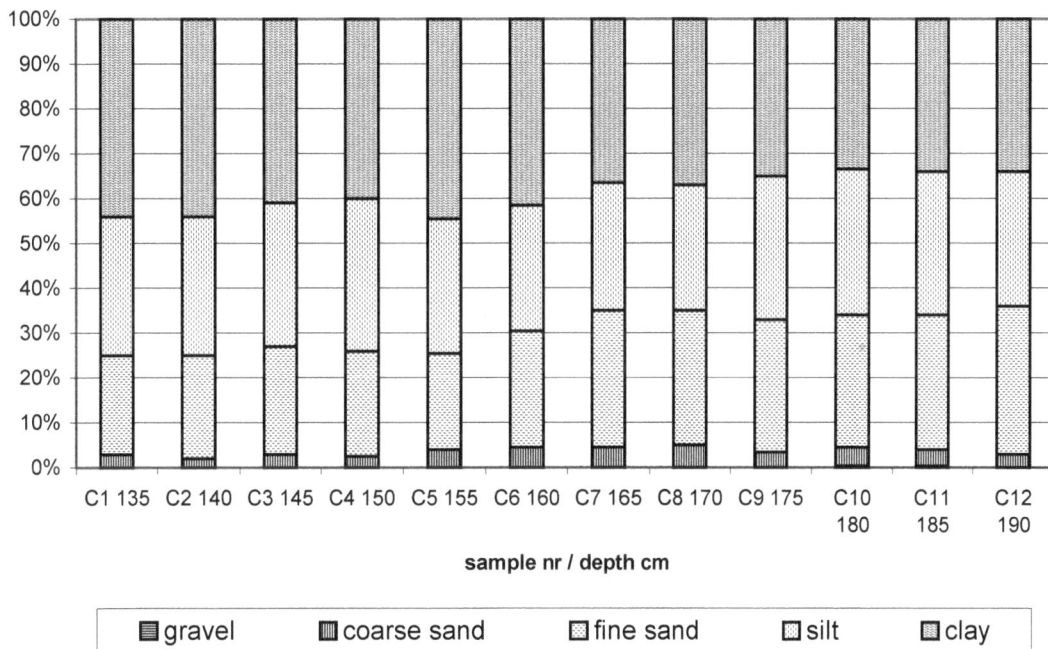

sample nr / depth cm

▨ gravel   ▥ coarse sand   ▦ fine sand   ▫ silt   ▨ clay

PALYNOLOGY (M.M.L. and M.M.S.)

Materials and method

Palynological analyses were carried out on sediments from three sequences in the site: A, B, and C (Fig. 1). The samples were collected from the layers chronologically antecedent, contemporary and subsequent to the inhabitation period of the site. The aim of the research was to reconstruct the floral context before, during and after the human presence.

The soil samples (3g) were treated with cold HCl, HF, hot HCl and NaOH, sieved (250 μm meshes), and included in a water/glycerol solution 50% v/v according to Bertolani & Marchetti (1960) with slight modifications (Arobba, 1986). Some samples were also treated with acetolytic method

(Erdtman, 1960). Observations were carried out using a light microscope.

The grains were identified with the aid of the literature (MOORE *et Alii*, 1991; REILLE, 1992; 1995) and the pollen reference collections (University of Florence). Pollen concentration (APF = Absolute Pollen Frequency) was calculated as the number of grains per gram according to Accorsi & Rodolfi (1975).

Results and discussion

All of the sediments proved to be very rich in grains, with the APF ranging between 2550 and 14800 grains per gram. The study sequences showed a partial overlapping. In particular, the sequence A presents one anthopogenic layer in the lowest portion; sequence B presents two anthopogenic layers in the lowest portion (cm 195 and cm 185); the sequence C is composed entirely of anthropogenic layers.

On the basis of the pollen analyses, it is possible to identify three distinct, subsequent phases in the vegetational history of the site. These phases can be clearly seen in the sequence B (Fig. 3, *above*).

During the ***first phase***, immediately before and during the time-span when the site was inhabited, the arboreal cover may have been extremely poor. In fact, the percentages of grains from trees, shrubs, and climbers (AP) are very low. *Pinus* is the only tree listed in the spectra with significant percentage values. Its grains are generally badly preserved; the few well-preserved grains are attributable to *P. sylvestris* L. In some levels, *Alnus* is also present. *Pinus* and *Alnus* are known to produce a large amount of pollen grains which are dispersed over a long-distance. Consequently, they are generally over-represented in the pollen spectra. *Salix* pollen grains appear in the two lowest samples. The herbs (NAP) are numerous. They include plants which grow in wet environments, such as Cyperaceae, *Typha,* and those plants which present *Sparganium* pollen type. Together with *Alnus*

and *Salix*; they attest to the presence of swamps or water bodies near the site. The anthropogenic indicators, which include Urticaceae, are scant in the spectra.

The ***second phase*** corresponds to the period immediately after the abandoning of the site. The AP appear to have increased; their percentages are always below 60%. They are still represented by *Pinus* and *Alnus,* as well as by *Quercus* and sporadic appearances of *Betula, Tilia, Salix,* and Cupressaceae. NAP are still numerous; most of them belong to genera/families which are common in open areas, such as grasslands or meadows. The results suggest a period characterised by a higher dryness than the previous one.

The ***third phase*** is characterised by a higher percentage of AP than in the previous phases: in fact, all of the samples present AP above 60%. Many trees in the list are the ones that grow on the mountain reliefs today, such as *Picea, Abies, Fagus.* The mixed oak forest is well-represented, composed mainly of *Quercus, Tilia, Corylus* and *Acer.* Probably, a mesohygrouphilous forest, with *Alnus* and *Salix*, was probably widespread in the valley bottom. *Castanea, Juglans* and *Vitis* appear in the most recent samples. These plants were probably cultivated or, at least, looked after by man.

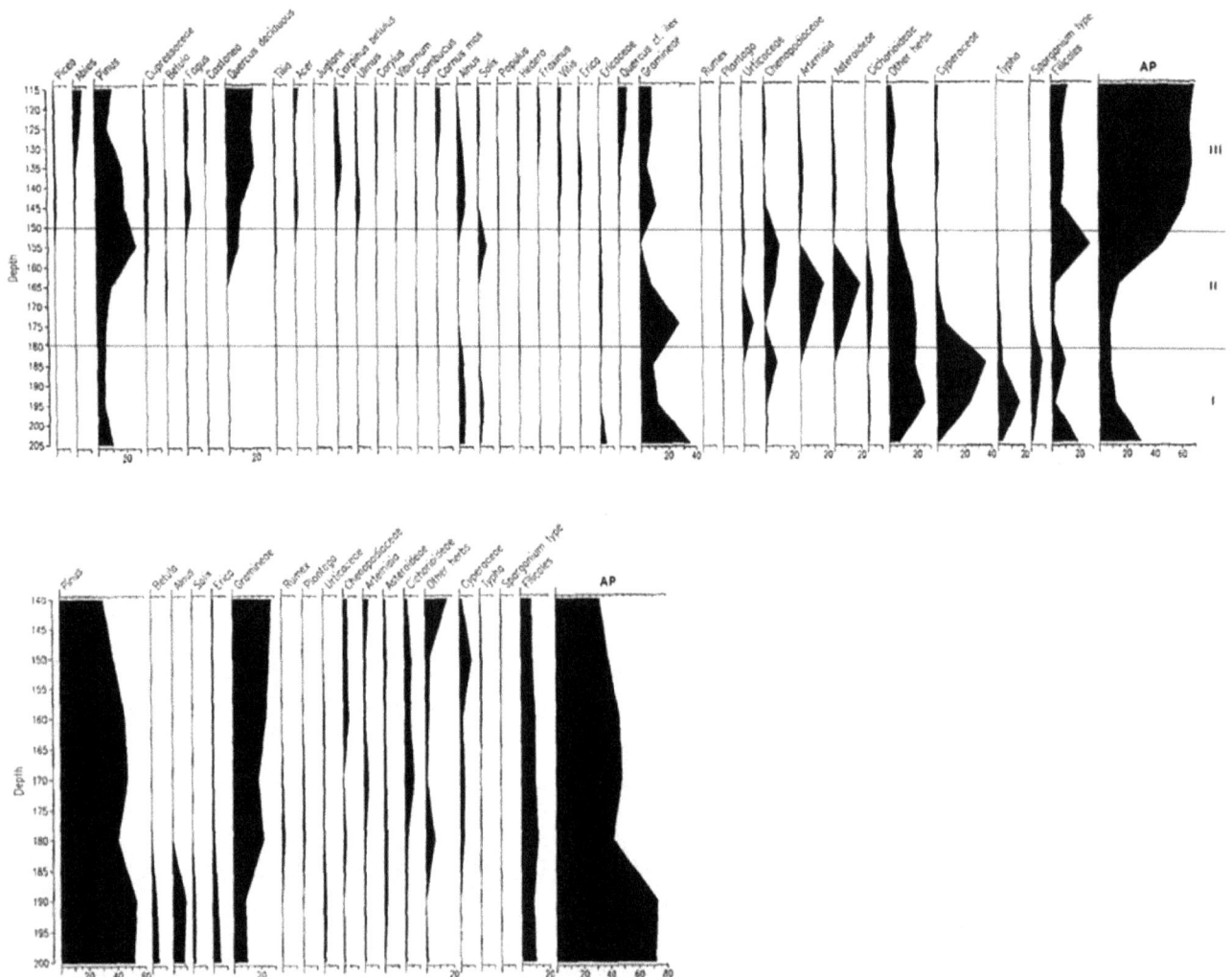

Fig. 3 - *above:* Pollen diagram of the sequence B. I=first phase; II=second phase; III=third phase.
below: Pollen diagram of the sequence C.

In this sequence it is possible to note a general trend towards a climatic improvement in the upper part of the sequence. This improvement concerned the temperatures and it was accompanied by an increasing dryness towards the most recent samples. In the last phase, the forest cover became remarkable; it was quite similar to the one we can observe today in the same area.

The results emerging from the analysis of the sequence B are substantially confirmed by sequence A. However, the latter sequence was shorter than the A and was discontinuous.

The sequence C (Fig. 3, *below*) presents a wider and more detailed picture of the human inhabitation period than the sequences A and B. The entire period is characterised by low percentages of AP. These are represented by *Pinus* (*P. sylvestris* p.p.) which presents decreasing percentages as we move upwards. In the two lowest samples, other trees are present: *Betula*, a microthermic plant; *Alnus* and *Salix*, trees which grow in wet environments. The poor arboreal cover, characterized by the presence of *P. sylvestris* and *Betula*, attests to a rather cold climate.

In the spectra, some herbs result as increasing, in particular Gramineae, Chenopodiaceaea, and Compositae, these last are divided into *Artemisia*, Asteroideae, and Cichorioideae in the diagram.

The early decrease in the hygrophilous *Alnus* and *Salix* is not followed a the decrease in the hygrophilous and hydrophilous herbs such as Cyperaceae and *Typha*. However, in the uppermost samples, all the hygrophilous and hydrophilous plants are scant, while the increased percentages of Chenopodiaceae and *Artemisia* attest to a trend towards dryness. Anthropogenic indicators are scant.

CHARCOAL ANALYSIS OF BILANCINO HEARTH Q36 (G. G., S. P.)

Samples of the charcoal remains, which showed an almost uniform distribution within the structure, were taken from the hearth Q 36. This hearth was built with sandstones: the stones were placed in a semicircle shape in the first phase of utilization, later on, other stones were placed on the top to cover the structure when it was not in use (Aranguren *et Alii*, 2001).

Some samples were collected from among the cover stones and others from among the stones of the hearth (Fig. 1).

During sample preparation special care was taken when separating the very small charcoal pieces from the englobing dry-clay: a first attempt run in water resulted in the total disgregation of the charcoal, mechanical separation was therefore carried out using a bistoury blade and an optical microscope.

This resulted in the collection of 70 very small charcoal pieces, about 2-3 mm³, which were then cut to produce transversal, longitudinal tangential and longitudinal radial

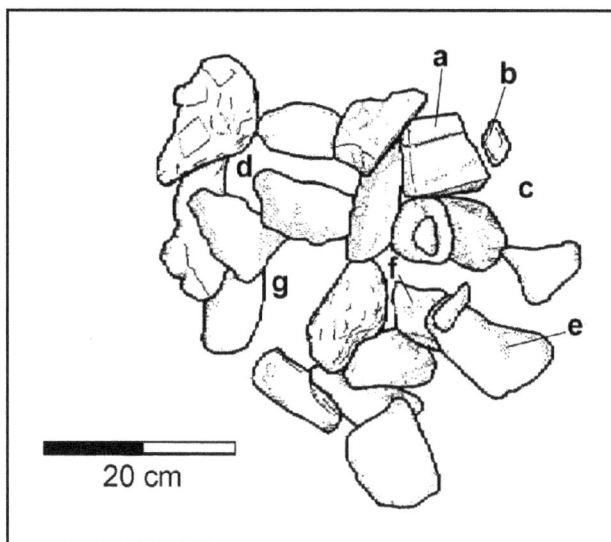

Fig. 4 – The structured hearth with spotted samplings.

sections. These were covered with a conducting golden film and observed with a scanning electron microscope (Philips 505).

Botanical species were identified by comparing the microscope images with ones in reference literature (Gellini *et Alii*, 1979; Schweingruber, 1990).

For some samples it was not possible to make any attribution because the wooden frame was completely destroyed by the fast combustion, as their gas-bubbled surface showed; for some charcoal we could only recognize the *Genus*, as it was not possible to make all the diagnostic sections.

The anatomic identification allowed to identify 41 samples of *Pinus sylvestris* L., 7 of *Fraxinus* cfr. *excelsior*, 1 of *Salix/Populus*[1] and 1 of *Erica* sp., as well as 7 hardwoods and 4 softwoods. The distribution of the species in the furnace samples is shown below in the Table:

Sample position	Botanical species	Number of fragments
a	Not identified	5
b	*Pinus sylvestris* L.	3
	Salix/Populus	1
	Hardwood	1
c	Not identified	4
d	*Pinus sylvestris* L.	1
e	*Pinus sylvestris* L.	17
	Fraxinus cfr. *excelsior*	1
	Hardwood	3
f	*Pinus sylvestris* L.	7
	Softwood	4
g	*Pinus sylvestris* L.	13
	Fraxinus cfr. *excelsior*	6
	Erica sp.	1
	Hardwood	3

[1] It was not possible, for sample degradation, to distinguish between this species.

The most frequent species is *Pinus sylvestris* found throughout the furnace and especially in the hearth.

The number of fragments of *Pinus*, as those of the other species, may depend on the original abundance of wood as well as on the breaking of charcoal through time or during sample preparation. Charcoal findings were recovered after several accidental events due to the combustion conditions (oxygen supplying, charcoal brittleness, and so on) and moreover, the conifer wood, because of its resin content, burns quickly, with light flame, usually leaving only ashes (Castelletti, 1990).

In any case throughout the ages the wood used in a hearth could have been chosen either because it was easily found in the neighbourhood or because it had a good combustion yield.

Thus the results of the anthracological analysis are only representative of a small part of the floral context of an area (Castelletti, 1990) and, when possible, should be compared with pollen and carpological data (JONES, 1991; Eschenlor & Serneels, 1991; ABBATE EDLMANN *et Alii*, 1993; ABBATE EDLMANN *et Alii*, 1997).

The presence in the Bilancino charcoals of *Fraxinus* and *Salix/Populus*, alluvial terrain plants, with the last two living in wet land, points in favour of a landscape with water bodies.

It also appears that the last furnace supply was composed mainly of *Pinus sylvestris*, thus suggesting a wood gathering in the surroundings, where climate would have been cold.

CONCLUSION
(B.A., G.G. - M.M.L. - M.M.S.- S.P. - A.R. - G.R.)

Sedimentological and archaeobotanical data suggested that a wet environment was present around the Gravettian site.

The site was characterized by the closeness of water bodies, as indicated by the hydromorphic conditions in both the soil and alluvial-colluvial deposits, and by the presence of significant amount of hygrouphylous plants pollen grains (*Alnus, Salix, Typha, Sparganium* pollen type and Cyperaceae) and the charcoals (*Salix/Populus*). Probably, just this environmental condition dominated by the presence of water could have attracted prehistoric people, because of the abundance of natural resources in the wet areas. It also appears that the last hearth supply was composed mainly of Pinus sylvestris, thus suggesting a wood gathering in the surroundings, where the climate would have been cold.

The scheme inferred from both the superficial survey and the excavations, induces us to think to a temporary man's presence, due to the recurrence of flood events, as well as to mudflows from the upper slopes. The chaotic arrangement of the findings in the area closest to the footslope seems to strengthen this hypotesis.

Therefore, the Gravettian settlement of Bilancino was a seasonal camp, because during the rainy seasons the site was not suitable for settling. Probably, this is the reason for the small amounts of anthropogenic indicators (i.e. *Urticaceae*) in the pollen content of the samples.

The rich framework offered by sedimentological, palynological and anthracological data also confirms the functional interpretation of the settlement just suggested on first studies: the site of Bilancino was probably a seasonal camp for the harvesting and the processing of hygrophilous herbs, in particular of *Typha*. For these activities the inhabitants of Bilancino settlement needed a very special type of instrument that may be the Noailles burin: a small tool apt to engrave carefully weak material.

During the human settlement, the environmental situation, like the typological characteristic of the Bilancino lithic industry, is probably comparable with the ones resulting from some French Palaeolithic sites of the same age. This has been dated about 25.000 y. BP. Those sites have been attributed to the Tursac interstade (Würm III, phase VII). As during the French Tursac (Laville *et Alii*, 1983), our analyses showed that the forest cover was scarce. *Pinus* was dominant among the AP, in particular, *P. sylvestris*. Grains of the latter were identified and its wood was gathered and used for hearth.

The pollen analyses showed some changes in the floral composition and this let us to suppose that drastic changes occurred in the vegetational cover of the area surrounding the sampling place, during the last 25.000 years. In fact, the sequences indicate a progressive change toward modern forest vegetation, starting from the glacial microthermic pinewoods composed of *P. sylvestris*.

Author's addresses

Biancamaria ARANGUREN, Gianna GIACHI
Soprintendenza Archeologica per la Toscana, Via della Pergola 65 - 50121 Firenze, ITALY

Marta MARIOTTI LIPPI, Miria MORI SECCI
Dipartimento di Biologia Vegetale - Università di Firenze - Via G. La Pira, 4.- 50100 Firenze, ITALY

Stefano PACI, Anna REVEDIN
c/o Istituto Italiano di Preistoria e Protostoria – Via S. Egidio, 21 – 50122 Firenze, ITALY

Giuliano RODOLFI
Dipartimento di Scienze del Suolo , Università di Firenze – Piazzale delle Cascine, 15 – 50100 Firenze, ITALY.

Bibliography

ABBATE EDLMANN, M. L., BARGELLI, S., & GIACHI, G., 1993, Indagine antracologica. In: *Grotta della Serratura a Marina di Camerata*, edited by F. Martini. Città di Castello (PG): Garlatti e Razzai, p.47-53.

ABBATE EDLMANN, M. L., GIACHI, G. & MARIOTTI, M., 1997, Indagine antracologica. In: *Querciola*, edited by L. Sarti. Città di Castello (PG): Garlatti e Razzai, p. 33-39.

ACCORSI, C.A. & RODOLFI, G., 1975, Primi risultati sullo studio di un suolo calcimorfo delle Alpi Apuane in relazione ad analisi palinologiche e microbiologiche. *Boll. Soc. Ital. Scienze Suolo* 9, p. 35-51.

ARANGUREN, B. & REVEDIN, A., 1997, Il ciottolo inciso e utilizzato dall'insediamento gravettiano di Bilancino e i "ciottoli a cuppelle" in Italia. *Rivista di Scienze Preistoriche* XLVIII, p.187-222.

ARANGUREN, B. & REVEDIN, A., 1998, L'habitat gravettien de Bilancino (Barberino di Mugello - Italie centrale). *Proceedings of XIII International Congress of UISPP*, Forlì (Italy) 8-14 September 1996, vol. 2, sez. 6, p. 511-516.

ARANGUREN, B. & REVEDIN, A., 2001, Interprétation fonctionnelle d'un site gravettien à burins de Noailles, *L'Anthropologie* 105, p. 533-545.

ARANGUREN, B., GIACHI, G., PALLECCHI, P. & REVEDIN, A., 2001, Primi dati sul focolare gravettiano di Bilancino. In *Atti della XXXIV Riunione Scientifica IIPP "Preistoria e Protostoria della Toscana"*, Firenze: IIPP, 2001.

AROBBA, D., 1986, Tecniche palinologiche di laboratorio. *Boll. Acc. Gioenia Sci. Nat.* 19, p. 273-288.

BENVENUTI, M., 1995, Il controllo strutturale nei bacini intermontani plio-pleistocenici dell'Appennino Settentrionale: l'esempio della successione fluvio-lacustre del Mugello (Firenze). *Il Quaternario* 8 (1), p. 53-60.

BERTOLANI MARCHETTI, D., 1960, Metodo di preparazione dei sedimenti per l'analisi pollinica. *Atti Soc. Nat. Mat. di Modena* 91, p. 58-59.

CASTELLETTI, L., 1990, Legni e carboni in archeologia. In: *Scienze in archeologia. II Ciclo di lezioni di ricerca applicata in archeologia. Certosa di Passignano (Siena) 7-19 novembre 1988*, edited by T. Mannoni & A. Molinari. Firenze: All'Insegna del Giglio, p. 321-395.

ERDTMAN, G., 1960, The acetolysis method. A revised description. *Svensk Botanisk Tidskrift* 54 (4), p. 561-564.

LAVILLE, H., TURON, J.-L., TEXIER, J.-P., RAYNAL, J.-P., DELPECH, F., PAQUEREAU, M.-M., PRAT, F., DEBENATH, A., 1983, Histoire paleoclimatique de l'Aquitaine et du Golfe de Gascogne au Pleistocene superieur depuis le dernier Interglaciaire. *Bull. Inst. Géol. Bassin d'Aquitaine, Bordeaux* 34, p. 219-241.

MOORE, P.D., WEBB, J.A., COLLISON, M.E., 1991, *Pollen analysis*. Oxford: Blackwell Scientific Publications.

REILLE, M., 1992, *Pollen et spores d'Europe et d'Afrique du Nord*. Marseille: URA CNRS.

REILLE, M., 1995, *Pollen et spores d'Europe et d'Afrique du Nord. Supplement 1*. Marseille: URA CNRS.

RODOLFI, G., 1982, Considerazioni sulla evoluzione della rete idrografica nel Mugello. *Studi e Materiali* 5, p. 298-302.

RODOLFI, G., SAVIO, S. & MARTENS, P., 1978, Esperienze di cartografia tematica nel Mugello centrale (Firenze). *Ann. Ist. Sperim. Studio e difesa Suolo* 9, p. 67-138.

SANESI, G., 1965, Geologia e morfologia dell'antico bacino lacustre del Mugello. *Boll. Soc. Geol. It* 84, p. 170-252.

SCHWEINGRUBER, F. H., 1990, *Anatomie europäischer Hölzer. Anatomy of European Woods*. Bern und Stuttgart: Paul Haupt Verlag.

LA FAUNE DE LA FIN DU PLEISTOCENE DANS LA HAUTE VALLEE DE L'AUDE : L'EXEMPLE DE LA GROTTE DE CASTEL 2 A BESSEDE-DE-SAULT (AUDE, FRANCE)

Jacques PERNAUD, Jérôme QUILES & Florent RIVALS

Résumé : La grotte de Castel 2 se situe en moyenne montagne dans la haute vallée de l'Aude (France). Les fouilles ont permis la découverte de nombreux restes de carnivores, hyènes et ours des cavernes et de petits bovidés, surtout bouquetin des Pyrénées. Les différents ossements pemettent de considérer ce gisement comme datant de la fin du Pléistocène.

Abstract: The Castel 2 cave is found at a middle mountain altitude, in the high valley of the river Aude (France). The excavation yielded numerous carnivore remains (hyena and cave bear) and small bovid remains, especially *Capra pyrenaica*. The different bones show that the site may date of the end of Pleistocene.

Découverte en 1991 par les spéléologues Rémi Clamens et Raymond Pradelle, la grotte du Castel 2, à Bessède-de-Sault a fait l'objet de deux campagnes de recherches successives qui ont montré la grande richesse potentielle du site. La grotte se trouve dans la haute vallée de l'Aude, dans le sud de la France, à une altitude de 635 m soit déjà en moyenne montagne pyrénéenne.

La grotte est composée d'un boyau étroit et anguleux précédant un couloir menant à une grande salle. Ces deux dernières zones sont tapissées d'ossements plus ou moins concrétionnés.

Nous avons pratiqué un sondage dans le couloir et plusieurs sondages dans la grande salle. Dans tous les cas les ossements extraits sont nombreux et bien conservés. Ils proviennent essentiellement de deux groupes de mammifères les Carnivores et les Bovidés.

LES CARNIVORES

Plusieurs espèces de carnivores sont représentées dans la faune de Castel II. Certaines sont marginales comme le renard ou le blaireau. Deux sont plus abondantes : la hyène et l'ours.

L'HYENE DES CAVERNES

L'hyène des cavernes est représentée par des restes surtout crâniens (dents, hémi-mandibule). Les restes ont été découverts surtout dans la grande salle. Sur une hémi-mandibule gauche l'état d'usure des dents correspond à un individu adulte déjà dans la force de l'age (Michel, 2001). L'hyène est souvent présente dans les sites où l'ours est plus particulièrement important. Les traces des activités d'hyènes n'ont pas vraiment pu être identifiées : pas d'os sur lesquels les traces de dents soient nettes et aucun coprolithe. La grotte du Castel 2 peut toutefois permettre l'étude d'hyènes dans la partie orientale des Pyrénées (Clot et Duranthon, 1990)

LES URSIDAE DE CASTEL 2

Les fouilles de Castel 2 ont mis au jour 86 restes osseux rapportés à la famille des *Ursidae*. 67 ont pu être déterminé en *Ursus spelaeus* Rosenmüller & Heinroth, 1794, soit près de 78% du NR, les 19 autres n'ayant qu'une attribution générique. Sur les fémurs, le calcul du NMIc révèle la présence d'au moins 5 individus de l'espèce *U. spelaeus*.

A) Paléontologie

La surface occlusale de la seule M3 inf retrouvée est fortement ridulée et riche en tubercules secondaires. Ces fortes dimensions (26,5 x 20 mm) la placent proche de l'*U. spelaeus* d'Espagne (Torres, 1988), du Sud-Ouest de la France (Prat & Thibault, 1976) ou de la grotte de la Carrière (Clot, 1980). La présence d'une prémolaire vestigiale (P3 sup) relève d'un certain « archaïsme » (Argant, 1991), mais fait exception. Les deux canines supérieures conservées montrent un gonflement mésio-distal racinaire très exagéré, caractéristique de l'ours des cavernes (Argant, 1991).

Le squelette post-céphalique très robuste possède d'importantes insertions musculaires. L'échantillon de population montre des caractères hyper-spéléens, notamment pour le carpe et le tarse, résultant d'une plantigradie accusée (Chagneau, 1985). Tibia et ulna ont une longueur moindre que fémur et humérus, atteignant un rapport de 1 sur 2 pour l'arrière train, vraisemblablement très surbaissé.

Les métapodes d'*U. spelaeus* et *U. deningeri* se révèlent plus robustes que leurs homologues de l'espèce arctos (Chagneau, 1985), ce qui est systématiquement le cas pour les 8 exemplaires de Castel 2 (2 Mc II, 1 Mt II, 3 Mc IV, et 2 Mc V). L'indice de la longueur totale en pourcentage du diamètre transversal médian est toujours équivalent à celui de l'*U. spelaeus* du S-O de la France. Les régions articulaires sont plus volumineuses pour les ours de la lignée spéléenne, l'extrémité distale étant particulièrement élargie (Chagneau, 1985). A Castel 2, le DT distal se situe toujours à la limite supérieure de l'intervalle de variation de l'ours des cavernes.

En vue latérale, la face dorsale est sub-rectiligne, contrairement à *U. arctos* où elle est légèrement arquée. La face palmaire ou plantaire possède des insertions musculaires en forme de fortes crêtes, jamais aussi développées chez *U. arctos* (Chagneau, 1985).

L'intervalle de variation ostéométrique des deux éléments les mieux représentés, fémurs et calcanéums, est comparable à celui d'une moyenne effectuée pour l'*Ursus spelaeus* du Sud-Ouest de la France (Prat & Thibault, 1976, tabl. 1 & 2). Ils se placent entre les valeurs relevées pour *U. spelaeus* aux Furtins et Plo del May pour les calcaneums, et aux Furtins et en Espagne pour les fémurs.

La corrélation des mesures ostéométriques aux observations morphologiques permet d'établir que le degré évolutif atteint par l'échantillon de population d'*U. spelaeus* de Castel 2, correspond à celui des grandes formes d'ours des cavernes de la fin du stade isotopique 4.

B) Archéozoologie

L'état de conservation général des restes est bon. Une seule connexion anatomique a été retrouvée, mais les ossements ne semblent pas avoir subi d'importants déplacements avant leur découverte.

La représentation squelettique suggère une conservation différentielle de l'assemblage, dont presque toutes les parties anatomiques ont été conservées, notamment le rachis (21% du NR). La présence de squelettes d'ours des cavernes sub-entiers in-situ est confirmée par un indice de représentativité (% NMI/NR = 7,5) faible.

Les jeunes individus sont nombreux, représentés par 18 dents lactéales diverses, un bourgeon de M3 inf et 7 os post-crânien, correspondant à 30% du NR. L'état de surface montre une absence d'actions anthropique ou carnivore. La fragmentation de l'assemblage est relativement faible, 40% des restes étant encore entiers. L'indice de fragmentation (% ossements entiers / ossements fragmentés = 62,3) s'explique principalement par les actions conjuguées de la calcification (8,2% du NR) et de la fissuration (7% du NR), consécutif du weathering et/ou de l'écrasement par le poids des sédiments.

L'assemblage osseux d'ours des cavernes de Castel 2 est d'origine intrusive. Sa présence dans la grotte résulte très probablement d'une mort naturelle pendant l'hivernation.

LES CAPRINES

Le chamois

Le genre *Rupicapra* est représenté par seulement deux restes (ramassés en surface dans le couloir) :

- une mandibule gauche portant les 3ᵉ et 4ᵉ prémolaires lactéales, et la deuxième molaire en train de sortir. Elle correspond à un jeune individu d'environ 1 an (selon les données de Silver, 1969).

- un tibia droit dont seule l'extrémité proximale est absente. L'extrémité distale est déjà bien soudée, cet individu est âgé de plus de 4 ans.

Ces deux ossements appartiennent donc à deux individus différents : un jeune et un adulte.

L'absence de chevilles osseuses rend impossible la détermination spécifique de ce chamois. Toutefois les données biométriques réalisées sur le tibia montrent qu'il s'agit d'un animal de petite taille qui s'inscrit dans la gamme de variation des isards fossiles (Tableau 3).

Mesures du tibia de chamois du Castel 2 :

DT distal articulaire = 23,5 mm

DT distal maximum = 26,2 mm

DAP distal maximum = 20,8 mm

D'autres vestiges de chamois sont toujours présents dans la grotte, notamment dans la grande salle où un métapode a été observé en surface du remplissage.

Le bouquetin

A) Inventaire des restes

Le nombre de restes attribués au bouquetin s'élève à 204. Dans le détail, ont été identifiés :

- 1 neurocrâne avec les chevilles osseuses, 1 condyle occipital, 6 fragments crâniens, 1 os hyoïde ; 6 mandibules dont 4 avec les dents déciduales, des dents inférieures : 2 incisives lactéales et 4 définitives (1 I_1, 2 I_2, 1 I_3, 1 M_2 et 1 M_3) ; 5 maxillaires dont 3 portent les dents déciduales, des dents supérieures : 2 P^3, 2 P^4.

- 13 vertèbres cervicales dont 3 atlas et 1 axis, 25 vertèbres thoraciques, 9 vertèbres lombaires, 4 sternèbres et 44 côtes ou fragments de côtes.

- 3 scapulas, 5 humérus, 5 radio-ulna, 2 scaphoïdes, un semi-lunaire, un pyramidal, 2 métacarpes.

- 6 coxaux, 3 fémurs, 3 patellas, 6 tibias, 4 os malléolaires, 3 talus, 6 calcanéums, 3 cubonaviculaires, 3 grands cunéiformes, 2 petits cunéiformes, 10 métatarses.

- 7 premières phalanges, 5 deuxièmes phalanges et 3 troisièmes phalanges.

Tous les os du squelette sont bien représentés et bien conservés. Plusieurs restes découverts dans des zones voisines ont pu être associés. Ces associations concernent essentiellement les os des pattes et les vertèbres. Les ossements de bouquetins ont donc subi très peu de déplacements avant leur enfouissement.

B) Justification spécifique

- Les chevilles osseuses

Le crâne découvert en surface dans la grande salle correspond au neurocrâne. Les deux chevilles osseuses de

Tableau 1 : Dimensions ostéométriques des calcaneums d'*U. spelaeus* de Castel 2.

Calcaneum		M1 Lt	M2 DT max	M3 DDP max	M4 H manubrium	M5 angle axe os- tub lig calca	M6 (DTmax /Lt)x100	M7 (H manubrium/ Lt)x100	M8 (DDP/DT) x100
Ursus arctos									
Taubach	n	27	21				21		
Kurten (1977)	min	87	60						
	max	110	74						
mâle	m	100,6	65,4				65,01		
Taubach	n	12	10				10		
Kurten (1977)	min	83	56						
	max	99	63						
femelle	m	90,3	59,1				65,45		
Ursus deningeri									
Scharzfeld	n	22							
Schütt (1968)	min	79							
	max	110,4							
	m	86,4							
Nauterie c. 8-11	n	46	33	54		52	27	44	32
Prat & Thibault (1976)	min	79	52	29		23,3	60,22	41,48	67,92
	max	104	70	52		55,5	70	53,68	81,08
	m	91,27	59,56	43,44		36,05	64,46	49,04	74,22
Ursus spelaeus									
Azé I-3	n	21	12	27	22		12	21	12
Argant (1991)	min	71,5	41	35	36				
	max	87,9	60	46,2	50				
	m	78,5	52,79	40,29	41,11		67,25	52,37	76,32
	ic								
	s	4,9	4,7	2,6	3,8				
	v	6,3	9	6,4	9,4				
Moy S-O France	n	33					32		32
Prat & Thibault (1976)	min	82,5					62,7		61,4
	max	117,5					74,7		78,3
	m	98,25					69,19		69,77
	ic	3,77					1,07		1,57
	s	10,83					3,03		4,44
Furtins	n	49	31	63			31	1	31
Argant (1991)	min	81	51	41					
	max	112,3	73,8	56,2					
	m	99,71	65,33	48,45			65,52	41,12	74,16
	ic								
	s	7,9	6,9	3,4					
	v	7,9	10,5	7					
Castel 2	n	1	2	3	1	3		1	2
mes. pers.	min		54,8	41,5		33			
	max		70,3	52,9		38			
	m	**107**	**62,6**	**47**	**64,4**	**35,3**		**60,2**	**71,1**
	ic		10,7	5,3		2,3			
	s		7,8	4,7		2,1			
	v		60,1	21,7		4,2			
Plo-del-May	n	33	37	47		46	31	34	37
Laville & al. (1972)	min	86,5	45	32		37,8	64,04	42,99	57,34
	max	113	79,5	49		57,8	73,33	54,9	71,11
	m	96,56	65,76	41,97		45,56	68,25	48,93	64,34
	ic	3,02	2,5	1,09		1,35	0,88	0,92	1,07
	s	8,5	7,45	3,7		4,55	2,41	2,63	3,18

Tableau 2 : Dimensions ostéométriques des fémurs de Castel 2 par rapport à divers gisments à *Ursidae* pléistocène.

Fémur		M1 — Lt	M2 — DCC p	M3 — DT p	M4 — DCC m	M5 — DT m	M6 — DCC d	M7 — DT d	M8 — DT max tête	M8 — (DTm/Lt) x100	M9 — (DTd/Lt) x100
Ursus arctos											
Moy. Espagne	n	22		24		25		27	24	21	21
Torres (1988)	min	320,6		72,4		25,8		62	36,3	7	19
	max	470		112		42,2		97,8	55,6	10	23
	m	402,4		94,5		34,6		81,5	45,4	8,5	20,1
	ic										
	s	46,5		13,27		5,21		11,63	6,13	0,68	1,15
moy. Pays Basque	n	4		4		4		4	4		4
Altuna (1973)	min	320		75		26		62	35,5		19,2
	max	420		99		36		82	45,5		20,2
	m	372,3		87,37		30,75		73,37	40,75		19,5
Ursus deningeri											
Moy. Espagne	n	3	3	4		20		6	33		
Torres (1988)	min	373	88,6	92,5		34		73,1	43		
	max	433,5	105,4	106,4		40,4		100	62,1		
	m			100,7		36,4		83	48		
	ic										
	s		5,18			3,05		3,15	4,49		
Scharzfeld	n	7						17			7
Schütt (1968)	min	375						81			20,7
	max	447						107			25
	m	408						91			22,1
Nauterie c. 8-11	n	18						15		18	15
Prat & Thibault (1976)	min	372								9,2	20,3
	max	469								11	23,9
	m	422,8						93,55		9,96	22,08
	ic	16,4								0,29	0,66
	s	32,97								0,59	1,19
Ursus spelaeus											
Moy. Espagne	n	30		40		81		62	57	30	30
Torres (1988)	min	356,8		90		30		73	41	9	21
	max	527		150		55,5		128,8	68	12	25
	m	419,2		114,4		43,6		96	54	10	22,8
	ic										
	s	50,16		17,77		5,15		11,26	7,64	0,76	1,17
Moy S-O France	n	25						24		22	24
Prat & Thibault (1976)	min	374,5								9,7	21,6
	max	500								12,1	24,6
	m	448,9						105,7		10,55	23,54
	ic	15,24								0,26	0,35
	s	37								0,6	0,83
Furtins	n	3	7	8	9	9	11	10			
Argant (1991)	min	434,4	48	100,4	33	42,3	67,5	88			
	max	459	63,7	133,3	40	50	90	110,4			
	m	447,1	57,7	124,3	35,28	47,04	80,79	102,3			
	ic										
	s	10	4,5	10,4	2,1	2,7	6,2	5,9			
	v	2,2	7,8	8,3	6,1	5,7	7,7	5,7			
Castel 2	n	1	2	2	2	2		2	3	1	1
mes. pers.	min		56,4	125,5	31,8	49,3		104,2	55,3		
	max		61,4	135,7	33,7	49,9		114,3	59,9		
	m	**451,7**	**58,9**	**130,6**	**32,8**	**49,6**		**109,3**	57,9	**11,05**	**25,3**
	ic		3,5	7,1	1,3	0,4		7	2,2		
	s		6,3	26	0,9	0,1		25,5	3,7		
	v		2,5	5,1	1	0,3		5,1	1,9		

Tableau 3 – Diamètre transversal distal maximum du tibia de Chamois. Comparaisons

	Espèce	Minimum	Maximum	Moyenne	Référence
Les Cèdres	*R. rupicapra*			28,7	Crégut-Bonnoure, 1995
Nestier	*R. pyrenaica*			27,6	Clot et Marsan, 1986
La Vache	*R. pyrenaica*	27,5	30,0	29,1	Koby, 1964
Labastide	*R. pyrenaica*	25,5	27,7	26,6	Clot, 1988
Castel 2				26,2	

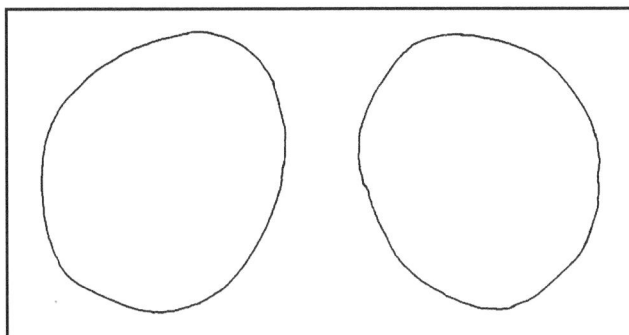

Figure 1 - Section basale des chevilles osseuses du bouquetin de Castel 2 (Echelle 1:1)

Dimensions du crâne du Castel 2 : (mm)
- Largeur maximale entre les deux condyles — 61,4
- Largeur maximale du foramen magnum — 24,9
- Hauteur du foramen magnum (basion - opisthion) — 18,9
- Largeur occipitale minimale (au niveau du sillon temporal) — 81,7
- Hauteur maximale de la région occipitale (basion - point le plus haut entre les deux chevilles osseuses) — 127,0
- Hauteur minimale de la région occipitale (opisthion - point le plus haut entre les deux chevilles osseuses) — 106,2
- Largeur entre les bords internes des bases des chevilles osseuses — 41,5
- Largeur entre les bords latéraux des bases des chevilles osseuses — 104,8

cornes sont entières, l'extrémité distale n'est pas altérée. En avant des cornes, il est fracturé au niveau des foramens supra-orbitaires, des arcades sourcilières et des arcades zygomatiques. La suture fronto-pariétale est quasiment rectiligne, la suture pariéto-occipitale décrit un angle. Cette disposition caractérise le genre *Capra* par rapport à l'*Ovis* chez lequel la disposition est inversée (Boessneck, 1969). Les os frontaux sont fortement pneumatisés. Le pariétal est bien développé comme chez tous les membres de la famille des Caprinae. Au niveau des deux os temporaux, la crête temporale est brisée dès la base, la bulle tympanique et le méat acoustique externe sont absents. Les rochers sont en place et bien conservés. Le condyle occipital est bien conservé mis à part les processus jugulaires qui sont cassés à la base.

Les chevilles osseuses sont relativement écartées l'une de l'autre à la base et présentent une courbure qui se combine à une torsion régulière (torsion hélicoïdale) vers l'extérieur. La section est ovalaire (Fig. 1). Toutes ces caractéristiques morphologiques des chevilles osseuses démontrent l'appartenance du bouquetin du Castel 2 à l'espèce *Capra pyrenaica*.

De part ses dimensions, le bouquetin du Castel 2 s'inscrit dans la gamme de variation des *Capra pyrenaica* actuels et fossiles. Compte tenu de la soudure des sutures crâniennes,

de la longueur des chevilles osseuses et de l'existence d'un dimorphisme sexuel très important observé sur les bouquetins actuels (Tableau 4), ce crâne semble être celui d'un individu femelle adulte.

- La troisième molaire supérieure

La troisième molaire supérieure présente des surfaces interstylaires de largeur sub-égale. Le métastyle est sub-vertical et assez épais à la base ; il forme une aile métastylaire peu développée. Cette dent présente toutes les caractéristiques qui permettent de justifier la détermination du bouquetin des Pyrénées, par rapport au bouquetin des Alpes (Crégut-Bonnoure, 1992).

C) Les individus

L'approche archéozoologique du matériel a été facilitée par un bon état de conservation des restes. Deux paramètres ont été déterminés, l'âge et le sexe des individus.

L'âge des animaux est estimé grâce à l'observation des stades d'éruption des dents lactéales et définitives sur les mandibules et l'étude du degré d'usure des dents ainsi que par l'analyse de la chronologie de soudure des os. Par cette méthode, chacun de ces restes a été attribué à une classe d'âge (jeunes, sub-adultes, adultes et adultes âgés).

Tableau 4 - Chevilles osseuses de Bouquetin. Comparaisons

Mesures en mm		Périmètre à la base	Diamètre antéro-postérieur	Diamètre Transversal
Capra pyrenaica Grotte du Castel 2	Droite	110,4	36,2	32,8
	Gauche	113,5	37,1	34,5
Capra pyrenaica Grotte de Malarnaud (Griggo, 1991)	Droite	222,5	74,4	65,5
	Gauche	221,5	71,3	63,5
Capra pyrenaica **Grotte de La Vache (Koby, 1958)**			65 à 67	
Capra ibex Grimaldi (Boule, 1910)		max : 260	max : 85	
Capra ibex Monaco (Chaix et Desse, 1982)	Mâles	188,0 - 270,0 X=248,4 (N=31)	59,2 – 92,5 x=83,5 (N=31)	49,2 - 76,4 x=67,3 (N=31)
	Femelles	62,0 - 140,0 X=114,0 (N=45)	25,5 – 48,5 x=39,6 (N=44)	21,4 - 38,7 x=31,5 (N=45)
Capra ibex Actuel (Couturier, 1962)	Mâles	166,0 - 262,0 X=216,0 (N=85)	58,9 – 88,0 x=71,0 (N=85)	51,0 - 74,0 x=62,2 (N=85)
	Femelles	89,0 - 112,0 X=99,9 (N=45)	30,0 – 43,0 x=42,0 (N=45)	24,0 - 31,0 x=27,5 (N=45)

Le sexe est déterminée sur l'os coxal, par une méthode uniquement basée sur la morphologie (Edwards *et al.*, 1982; Prummel et Frisch, 1986).

Ces deux paramètres ont été utilisés afin de calculer le nombre minimum d'individus par combinaison.

D'après le matériel étudié, le bouquetin est représenté par au moins 6 individus : 4 jeunes dont une femelle, 1 sub-adulte de sexe mâle et 1 femelle adulte.

Si le sexe n'apporte aucune information importante, le nombre de jeunes est relativement élevé. De plus, la corrélation avec les données de la répartition spatiale des restes et les associations réalisées, suggère un type d'accumulation accompli par des carnivores. Cette hypothèse va être prise en compte lors de l'analyse des traces sur les ossements.

CONCLUSION

Nous remarquons donc une grande abondance d'ossements dans les différentes zones de la grotte du Castel 2 à Bessède-de-Sault. Les deux espèces dominantes sont l'ours des cavernes *Ursus spelaeus* et le bouquetin des Pyrénées *Capra pyrenaïca*. L'hyène des cavernes *Crocuta spelaca* est plus marginale.

Les ossements sont en bon état de conservation. En particulier les restes de bouquetin des Pyrénées. Cela pose la question de la contemporainéïté des dépôts de bouquetins et de l'occupation par les grands carnivores. Les accumulations osseuses ne présentent pas de fracturation et des répartitions anatomiques particulières ce qui pourrait exclure le rôle de l'hyène dans leur dépôt (Tournepiche et al., 1995) Les vertèbres et les côtes sont présentes, les os de

petites tailles aussi. Le fait même que des os de bouquetin soient si bien conservés excluerait le rôle de l'hyène des cavernes (Guadelli, 1989). L'ours, de petite taille, toutes proportions gardées, pourrait être assez tardif. Il ne semble pas non plus être lié aux accumulations de restes de bouquetin, des os entiers et les vertèbres étant présentes (Tavoso et al., 1990).

Le grotte du Castel 2 est probablement un repaire de grands carnivores, surtout ours des cavernes mais aussi hyènes des cavernes. La plupart des restes de bouquetins, en très bon état de conservation, correspondent sans doute à un dépôt primaire pas forcément lié à l'action des carnivores, les traces de dents sont du reste très rares. Aucune activité humaine n'a été identifiée. Il est cependant intéressant de poursuivre l'étude de cette faune et de faire des comparaisons avec celles des sites préhistoriques locaux (Fontana, 1999).

Adresses des auteurs

J. PERNAUD
Musée de Tautavel
Centre Européen de Préhistoire
66720 Tautavel, FRANCE

J. QUILES
Institut de Paléontologie Humaine
et Laboratoire de Préhistoire du Muséum national d'Histoire naturelle
1, rue René Panhard
75013 Paris, FRANCE

F. RIVALS
Laboratoire de Préhistoire de l'Université de Perpignan
et
UMR 5590 du CNRS
66720 Tautavel, FRANCE

Bibliographie

ALTUNA J., 1973, Hallazgos de Oso Pardo (*Ursus arctos*, Mammalia) en cuevas del Pais Vasco. *Soc. Scein. Nat.*, 2 (2), p. 121-170, San Sebastian.

ARGANT A., 1991, *Carnivores quatrenaires de Bourgogne. Documents du Laboratoire de Géologie de Lyon*, Lyon.

BALLESIO R., 1983, Le Gisement pléistocène supérieur de la grotte de Jaurens à Nespouls (Corrèze) : les Carnivores (Mammalia Carnivora), III : Ursidae. *Nouv. Arch. Mus. Hist. Na t. Lyon*, 21, p. 9-43, Lyon.

BOESSNECK J., 1969, Osteological differences between sheep (*Ovis aries* Linné) and goat (*Capra hircus* Linné). In *Science and Archaelology- A survey in progress and research*. Sous la direction de D. Brothwell et E Higgs, p. 331-358, Londres.

BOULE M., 1910, *Les grotets de Grimaldi (Baoussé Roussé). Géologie et Paléontologie (suite).* 1 (3-4), p. 157-362, Imprimerie de Monaco.

CHAGNEAU J., 1985, Contribution à l'étude des os des extrémités des pattes d'*Ursus deningeri romeviensis*. Comparaison avec *Ursus arctos* et *Ursus spelaeus*. *Bull. Soc. Anthropo. S. O.*, 20 (2-3), p. 61-107.

CHAIX L. & DESSE J., 1982, Les Bouquetins de l'Observatoire (Monaco) et des Baoussé Roussé (Grimaldi, Italie. *Bull. Mus. Anthropo. Préhist. de Monaco*, 26, p. 41-74.

CLOT A., 1980, La grotte de la Carrière (Gerde, Hautes-Pyrénées) : Stratigraphie et Paléontologie des Carnivores. *Travaux Laboratoire de Géologie de Toulouse*, p. 167-288, Toulouse..

CLOT A., 1998, Faune magdalénienne de la grande grotte de Labastide (Hautes-Pyrénées, France). *Munibe*, 40, p. 21-44.

CLOT A. & DURANTHON F., 1990, *Les Mammifères fossiles du Quaternaire dans les Pyrénées*. 153 p., Toulouse.

CLOT A. & MARSAN G., 1986, La grotte du Cap de la Bielle à Nestier (Hautes-Pyrénées), fouilles de M. Debeaux, 1960. *Gallia Préhistoire*, 29 (1), p. 63-141, Paris.

COUTURIER M.-A., 1962, *Le Bouquetin des Alpes (Capra aegagrus ibex ibex L.).* 1564 p., Imprimerie Allier, Grenoble.

CREGUT-BONNOURE E., 1992, Intérêt biostratigraphique de la morphologie dentaire de *Capra* (Mammalia, Bovidea). *Ann. Zool. Fennici*, 28, p. 273-290.

CREGUT-BONNOURE E., 1995, Les grands Mammifères. In *Le Gisement paléolithique moyen de la grotte des Cèdres (Var), Doc. Archéol. Fr.*, 4, sous la direction d'A. Defleur et E. Crégut-Bonnoure, p. 54-143.

EDWARDS J. K., MARCHINTON R. L. & SMITH G. F., 1982, Pelvic Girdle criteria for sex determination of White-tailed deer. *Journal of Wildlife management*, 46, p. 544-547.

KURTEN B., 1977, Bären und Hyänenreste aus dem Pleistozän von Taubach *Quartärpaläonto*, 2, p. 361-378.

FONTANA L., 1999, Mobilité et subsistance au Magdalénien dans le bassin de l'Aude. *Bull. Soc. Préhist. Fr.*, 86 (2), p. 175-190, Paris.

GRIGGO C., 1991, Le bouquetin de Malarnaud (Ariège); implications paléobiogéographiques. *Quaternaire*, 2, p. 76-82.

GUADELLI J.-L., 1989, Etude taphonomique du repaire d'hyènes de Camiac (Gironde, France). Eléments de Comparaison entre un site naturel et un gisement préhistorique. *Bull. Assoc. Fr. Et. Quaternaire*, 38, t. 2, p. 91-100, Paris.

KOBY F.-E., 1958, Ostéologie de la chèvre fossile des Pyrénées (*Capra pyrenaica*, Schinz). *Eclogae Geologicae Helvetiae*, 51, p. 475-480.

KOBY F.-E., 1964, Ostéologie de *Rupicapra pyrenaica* d'après les restes de la caverne de la Vache. *Bull. Soc. Préhist. Ariège*, 19, p. 15-31.

LAVILLE H., PRAT F. & THIBAULT C., 1972, Un Gisement à faune du Pléistocène moyen : la grotte de L'Eglise à Cénac et Saint-Julien (Dordogne). *Quaternaria*, 16, p. 71-119.

MICHEL P., 2001, A Propos d'une grotte-repaire d'hyène des cavernes avec indice de présence humaine dans le niveaux würmiens : la grotte d'Unikoté à Iholdy (Pyrénées-Atlantiques). *Mémoire d'habilation à diriger les recherches*. Université de Perpignan.

PRAT F. & THIBAULT C., 1976, Le Gisement de la Nauterie à Romieu (GERS), fouilles de 1967 à 1973. *Mém. Mus Natn. Hist. Nat*, 82, Paris.

PRUMMEL W. & FRISCH H. J., 1986, A Guide for the distinctionof species, sex and body size in bones of sheep and goat. *Journal of Archaeology*.

ROSENMÜLLER J. C. & HEINROTH H., 1794, Quaedam de Osisibus Fossilibus Animalis Cuisusdam. *Historian eius et cognitionem accuratioram illustrantia*. Leipzig.

SCHÜTT G., 1968, Die Cromerzeitlichen Bären aus der Einhomhöle bi Scharzfeld. *Mitt. Geol. Inst. T. H. – Hanover*, 7, p. 1-121.

SILVER I. A., 1969, The ageing of domestics animals. In *Science in archaeology*, sous la direction de D. Brothwel et E. S. Higgs. Londres.

TAVOSO A., CREGUT-BPNNOURE E., GUERIN C., PERNAUD-ORLIAC J. & CAMMAS R., 1990, La Grotte de la Niche à Montmaurin (Haute-Garonne, France). Nouvelles données biostratigraphiques et approche taphonomique. *C. R. Acad. Sci. Paris*, t. 310, série II, p. 95-100, Paris.

TORRES PEREZ-HIDALGO T., 1980, El Oso de las cavernas (*Ursus spelaeus*) de los niveles inferiores de Ekain. In El yacimiento prehistorico de la cueva de Ekain (Deba, Guipuzcoa) sous la direction de J. Altuna et J. M. Merino, *Eusko Ikaskuntza Sociedad de Estudios Vascos*, 112 (1-2), p. 151-154, Toulouse.

TORRES PEREZ-HIDALGO T., 1988, Osos (Mammalia, Carnivora, Ursidae) del Pleistoceno de la Peninsula Iberica. *Boletin geologico y minero de Espana*, Madrid.

TOURNEPICHE J.-F., COUTURE C., GUADELLI J.-L. & MICHEL P., 1996, Les restes néandertaliens du Repaire d'hyènes de la grotte de Rochelot (Saint-Amant de Bonnieure, Charente, France). *C. R. Acad. Sci. Paris*, t. 322, série II A, p.429-435, Paris.

DISTRIBUTION DES GRANDS MAMMIFÈRES EN FRANCE AUX DEUX DERNIERS EXTRÊMES CLIMATIQUES (18 KA ET 8 KA)

Jean Philip BRUGAL, Anne BRIDAULT, Jean-Luc GUIADELLI & Jean-Denis VIGNE

Résumé. Une présentation de gisements français ayant fourni des faunes de grands mammifères est donnée pour les deux derniers extrêmes climatiques (Pléistocène supérieur et Holocène). La distribution et la composition des associations permettent d'apprécier la dynamique des peuplements sur le territoire français, de mettre en évidence des absence de taxons et des zones refuges, ainsi que d'apprécier les changements et disparitions des espèces pendant la période transitionnelle Pléistocène-Holocène.

Abstract. An overview of large mammal communities in France is given for the last two climatic extremes (LGM and Holocene). The range and diversity of faunal associations allow us to clarify the presence/absence of species and the existence of refugia as well as the change and disappearance of species during the Pleistocene-Holocene transition.

La France, par sa position géographique et sa topographie, représente un carrefour d'influences climatique et biologique particulières au cours du Quaternaire en Europe de l'Ouest. Succédant au dernier maximum glaciaire (LGM), une dynamique de réchauffement climatique s'amorce vers 16000 ans BP (Bard *et al.* 1990), marqué par la fonte des calottes polaires, la réduction des glaciers de montagne et la disparition des zones englacées (par ex., Massif Central), la remontée du niveau marin et le réchauffement des eaux marins de surface. Cette dynamique entraîne d'importantes modifications des territoires disponibles pour les communautés animales et les couverts végétaux.

Suivant des recherches récentes, la déglaciation est aujourd'hui conçue comme un «phénomène pulsé» qui s'achève par la disparition de l'inlandsis scandinave entre 9000 et 8000 ans BP (Magny 1995). La circulation des courants océaniques, modifiée par ces épisodes de fonte massive de glace, a généré de brusques retours au froid (*i.e.*, événements d'Heinrich, oscillations de Dansgaard-Oeschger), malgré une insolation très forte (par ex., Labeyrie & Jouzel 1999). L'évolution climatique de la période comprise entre 16000 et 10000 ans BP (Tardiglaciaire) est alors caractérisée par une succession d'oscillations froides et tempérées de durées et d'amplitudes inégales (Magny 1995). Les températures augmentent à nouveau brutalement à partir de 10000 ans BP d'environ 2,8°C par siècle, marquant le début du dernier interglaciaire. Ce processus de réchauffement climatique s'accompagne d'une recolonisation de la végétation, d'abord toundroïde ou steppique, puis une reconquête arborée s'amorce dès le Bölling, entrecoupée de phases régressives, pour aboutir à l'installation durable de forêts tempérées de feuillus sur l'ensemble de l'Europe durant l'Holocène ancien (*i.e.*, Huntley & Birks 1981).

En ce qui concerne les peuplements mammaliens, nos connaissances sont encore très fragmentaires. La restitution des scénarios est complexe du fait de la grande mobilité potentielle des animaux et des facteurs biotiques qui ont fortement varié durant cette période. Pendant le maximum glaciaire, l'Europe de l'Ouest, véritable péninsule du continent eurasiatique, se retrouvait plus ou moins isolée, limitant de ce fait les échanges fauniques avec l'Est. L'Angleterre n'était alors séparée du continent que par une vaste plaine englacée. En Méditerranée, le bloc corso-sarde était proche du rivage italien. La mise en place du climat méditerranéen au début du Quaternaire joue également un rôle dans les peuplements mammaliens et s'oppose aux régions du nord plus homogènes structurellement (plaines).

La France, par ses reliefs et son hydrographie, a présenté, durant certaines époques du Quaternaire, des cloisonnements géographiques qui ont modelé une répartition en mosaïque des communautés de grands mammifères ; les espèces les plus sensibles au froid étant principalement réparties dans les zones méridionales. C'est probablement durant la courte période comprise entre le retrait des glaces et les transgressions marines, que les communautés animales ont pu très largement circuler au sein de régions entières alors ouvertes. Loin de pouvoir aujourd'hui restituer précisément la dynamique de ces mouvements migratoires à l'échelle de la France, l'objet de cet article est de rendre compte d'un état des peuplements mammaliens pour les deux extrêmes climatiques.

MATÉRIEL ET MÉTHODE

Dans le cadre de deux projets[1], l'un portant sur la cartographie des deux derniers extrêmes climatiques (Brulhet & Petit-Maire 1999 ; Petit-Maire 1999[2]) et l'autre sur la réalisation d'un Atlas diachroniques des mammifères de France, nous avons crée une base de données spécifique concernant les

---

[1]   Ces recherches s'appuient sur un groupe de travail du CNF-INQUA (resp. N. Petit-Maire, UMR 6636) et sur le programme «Dynamique de la Biodiversité et Environnement : Atlas diachronique des mammifères en France» (resp. J.-D. Vigne, ESA 8045).

[2]   Voir également les cartes sur le site http://www.cnrs.fr/dossiers/dosclim/ biblio/pigb12/08_mammiferes.htm.

grands et moyens mammifères (depuis la tailles des Lagomorphes jusqu'aux Proboscidiens) pour les gisements datés (C14). Une présentation des associations mammaliennes, respectivement à 18 ka et 8 ka, permet ensuite d'appréhender les variations taxonomiques (apparition, disparition) entre ces deux extrêmes sur le territoire français.

Seuls les grands mammifères ont été considérés : ils indiquent généralement le mésoenvironnement alors que les petits mammifères (rongeurs) reflètent les biotopes plus immédiats du site mais sont de meilleurs indicateurs des variations paléoclimatiques[3]. De plus, la microfaune est plus dépendante des protocoles de récolte et des conditions de préservation des ossements dans le sol et, elle n'est pas aussi fréquemment documentée dans les sites archéologiques qui constituent l'essentiel de notre corpus documentaire. A ce propos, les plus grandes espèces sont surtout indicatrices de biotopes exploités par les chasseurs préhistoriques. Les archéofaunes ne sont en effet pas forcément représentatives de la biodiversité locale, mais d'un sous-ensemble de celle-ci, que l'homme a choisi d'exploiter selon des modalités variées. Les chasseurs, eux-mêmes contraints, à un certain degré, par les conditions environnementales, ont pu en effet adopter des stratégies plus ou moins spécialisées selon les périodes (i.e., Bridault, 1997b), voire des options différentes selon les saisons, choix qui se reflètent dans la structure des spectres fauniques (une espèce dominant un spectre faunique, par exemple).

Chaque espèce au sein d'une association a été considérée en termes de présence-absence bien que son abondance relative puisse être prise ne compte. Il est par ailleurs bien connu que certains herbivores seront sur-représentés en raison de leur taille et de leur éthologie (troupeau par exemple : certains sites renferment jusqu'à 95% de restes de renne) au détriment d'autres espèces de gros herbivores et de carnivores. Cependant le regroupement géographique de plusieurs sites permet en général d'obtenir un spectre représentatif des faunes à un niveau régional pour une période donnée.

LE DERNIER MAXIMUM GLACIAIRE
(18 000 +/- 1000 ans BP)

L'environnement physique (Morzadec-Kerfoun, Van Vliet-Lanoë, Antoine in Petit-Maire, 1999) est caractérisé par la présence marquée de pergélisols continus (en altitude et au nord de la France) ou discontinu (notamment le long de la façade atlantique), de glaciers ainsi que d'importantes formations éoliennes (loess, sables). Dans ce contexte, le territoire est marqué par une importante baisse du niveau marin, -120 m par exemple en Méditerranée, augmentant considérablement l'étendue des plaines littorales. Le sud et sud-est de la France sont plus arides et les données paléobotaniques (Renault-Miskoscky et Girard in Petit-

Maire, 1999) indiquent des milieux contrastés avec trois formations végétales : forêt boréale, steppe boisée et steppe. Des îlots de boisements dispersés sont présents dans des zones refuges (ex. Dordogne).

Les données concernant les faunes du Pléistocène supérieur, en particulier du dernier glaciaire, sont très nombreuses en France. Elles proviennent principalement de restes fossiles découverts en contexte archéologique, correspondant aux cultures de la fin du Solutréen et du début du Magdalénien (Roblin-Jouve in Petit-Maire, 1999). L'attribution chronologique est fournie tout d'abord par leur appartenance culturelle étayée par des datations radiocarbones (charbons, os). Les attributions obtenues ont leurs propres limites : la première recouvrant des faciès industriels parfois bien différenciés ne pouvant par précisément servir de marqueur chronologique, et l'autre pouvant également être entachée d'erreur ; la confrontation entre ces deux approches est parfois intéressante.

Nous avons uniquement considéré les gisements présentant des données absolues (en réduisant d'autant leur nombre potentiel) et adopté un intervalle de +/- 1000 ans. Considérer ensemble les associations animales entre 16 000 et 20 000 ans (soit +/- 2000 ans) aurait entraîné un mélange de faunes, et donc une distorsion, en raison d'oscillations climatiques de plus ou moins fortes amplitudes («Interstades» de Laugerie vers 20-19 000 ans et de Lascaux, vers 17-16 000 ans) qui ont pu modifier significativement les aires de répartition des espèces. Cependant, en raison de la rareté des donnés dans le Nord de la France, deux exceptions ont été faites : pour les localités d'Hallines (présence d'un mammouth M.primigenius, daté de 16 000 +/-300 ans, Fagnart, 1988) et du Transloy (bison des steppes daté de 21 490+/-270 ans, Antoine, 199, in litt.). Ces deux espèces sont de bons marqueurs des conditions de froid régnant alors dans ces zones pour la période du dernier maximum glaciaire.

Trente gisements livrant des restes de grands mammifères ont été retenus pour la tranche de temps considérée. S'y ajoute le cas particulier de la grotte pariétale Cosquer à Marseille (Bouches-du-Rhône). Les dates ont été obtenues sur les peintures en ce qui concerne le bison, le cheval et le félin (cf. lion des cavernes) et nous avons rajouté les représentations gravées : mégacéros, pingouin (Pinguinus impennis, D'Errico, 1994) et phoque, attribuées à la même phase d'occupation.

La plupart de ces sites (87%) se trouvent en contexte archéologique ; certains (27%) ont aussi livré des données palynologiques. Il n'existe pas de gisements connus dans la partie Nord de la France (au-delà du 47°N). Cependant de très rares sites sont connus en Belgique et en Angleterre et on peut facilement imaginer une association de type arctique (nombreux rennes, chevaux et bison, mammouths) évoluant sur les prairies glacées du Bassin Parisien. Pour la Corse, les données issues du site de Castiglione indiquent que la faune de cette époque était composée de trois taxons seulement de grands mammifères, tous endémiques, un mégacéros nain Megaloceros (Nezoleipoceros cazioti), un canidé (cuon sarde) Cynotherium sardous et un lagomorphe (lapin-rat) Prolagus sardus (Salotti et al., 2000).

---

[3] Il reste toutefois que la microfaune est un excellent indicateur climatique avec les exemples de la présence d'espèce froise et/ou sèche durant le LGM, comme le lemming (par ex., Les Cottiers) et le spermophile ou Citelle (par ex., Les Cottiers, Laugerie Haute).

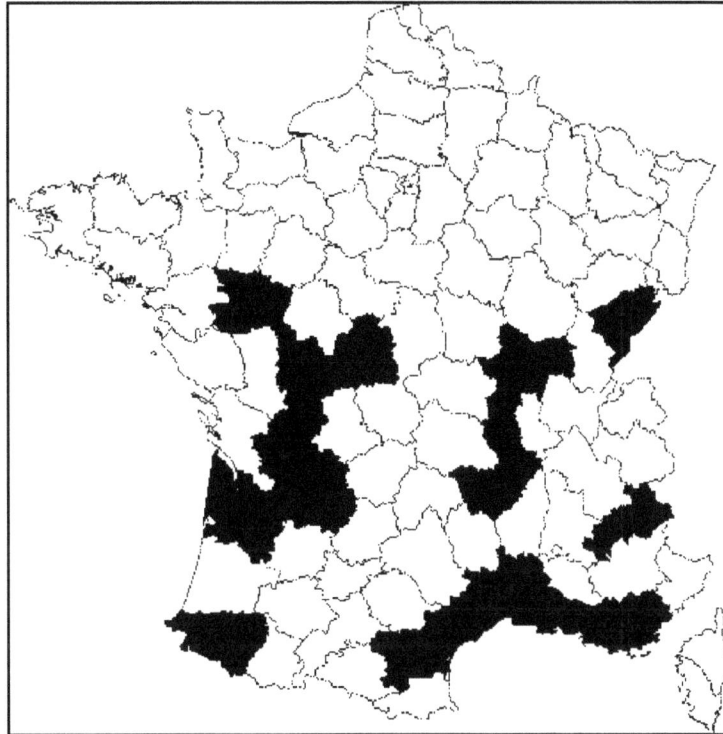

Figure 1. Répartition par département des mentions de grande faune à 18000 ± 1000 ans BP.
Abri Fritch (Inde), Roc de Sers (Vienne), Le Placard (Charente), Grotte de Rigney (Doubs), Solutré (Saône-et-Loire), Les Cottiers, Rond du Barry (Haute Loire), Saut de Perron (Loire), St Germain-la-Rivière, Roc de Marcamps (Gironde), Les Jamblancs, Combe Saunière, Lascaux, Laugerie-Haute, Abri Pataud, Abri Casserole (Dordogne), Le Piage, Barrières, Mude, Les Peyruges, Pégourié, Cuzoul de Vers (Lot), Oillascoa (Pyrénées Atlantiques), Lassac (Aude), La Roque 2 (Hérault), Chabot, Oullins, La Salpétriere (Gard), Cosquer (Bouches-du-Rhône), Le Pignon (Hautes-Alpes), Raynaude (Var).

On constate une concentration de gisements en quelques régions, certaines très denses (fig.1) : le grand Sud-Ouest, en particulier, regroupe prés de la moitié de la documentation. On peut distinguer 5 grandes zones : Pays de Loire et Poitou-Charente ; Rhône-Alpes et Massif Central ; Grand Sud-Ouest ; Languedoc-Roussillon et Provence. La composition taxinomique parait bonne avec une trentaine d'espèces. Le problème de mélange de couches dans les gisements stratifiés paléolithiques reste cependant posé : par exemple, la seule mention dans notre corpus du lynx, animal plutôt forestier, dans le site de St-Germain-la-Rivière (Gironde). Or dans ce site, une association particulièrement froide est présente avec l'antilope saïga dominante, accompagnée de renne, cheval, bison, loup et renard polaire ; par ailleurs, des lynx sont encore présents dans les zones froides (forêt boréale) du nord de l'Europe.

Principales caractéristiques de la grande faune (tabl. 1)

Les mammifères du dernier maximum glaciaire regroupent 13 espèces d'Herbivores, 2 de Lagomorphes et 9 espèces de Carnivores. Les sites naturels (n=4) contiennent généralement peu de taxons (1 à 3) alors que les sites préhistoriques sont plus riches (de 1 à 14 taxons, avec une moyenne générale de 7,2). Si on considère la densité des espèces selon les régions (17 départements sur 96), on obtient :

* Pays de Loire+Poitou- Charente (Indre, Vienne, Charente) : (n=3 sites) moyenne de 9,3 taxons, mais les sites sont très dispersés;

* Rhône-Alpes + Massif central (Doubs, Saône-et-Loire, Haute Loire, Loire) : (n=4) moyenne de 8,2 taxons ;

* Grand Sud-Ouest (Gironde, Dordogne, Lot, Pyrénées atlantiques) : (n=14) moyenne de 7,3 avec la distinction pour la Dordogne (n=6, X=7,8) et pour le Lot (n=6, X= 6,8) ;

* Languedoc-Roussillon (Aude, Hérault, Gard) : (n=5) 3,6

* Provence (Var, Hautes Alpes) : (n=2, sans Cosquer) : respectivement 3 (Le Pignon) et 5 (Rainaude) espèces, soit 4 en moyenne.

Les deux dernières régions possèdent une mauvaise représentation taxinomique, et en Provence par exemple aucun carnivore n'est signalé.

Selon leur fréquence, quatre groupes d'herbivores sont reconnus : tout d'abord une association typique du dernier glaciaire avec des espèces présentes sur tout le territoire : le renne (dans 87% des sites), cheval (83%) et grands Bovidés (essentiellement bison, 63%). Un deuxième groupe concerne le bouquetin, le chamois et le cerf élaphe qui restent des éléments assez communs et finalement un groupe comprenant l'antilope saïga, le chevreuil et le sanglier qui sont assez courants. Le mammouth, l'aurochs, le mégacéros et le cheval hydruntin sont des taxons assez rares. A ce sujet, l'introduction dans un site de fragment d'ivoire de mammouth, utilisé à des fins techniques ou symboliques

Tableau 1. Liste des espèces présentes et absentes en France à 18 et 8 ka (en gras les taxons les plus abondants). Rangement selon ordre d'importance pour 18 ka. X = disparition des taxons entre les deux extrêmes climatiques (entre parenthèses : taxons rares et double X : extinction). Cinq espèces domestiques se rajoutent à 8 ka (Ovis aries, Capra hircus, Bos taurus, Sus domesticus, Canis familiaris).

	18 +/- 1 Ka 29 taxa			8 +/- 1 Ka 20 taxa
1	***Rangifer tarandus***	renne	X	
	Equus caballus	cheval	X	*Equus* sp.
	Bison priscus	bison	(X)	(*Bison bonasus*)
2	*Capra ibex*	bouquetin		*Capra ibex*
	Rupicapra rupicapra	chamois		*Rupicapra rupicapra*
	Cervus elaphus	cerf		***Cervus elaphus***
3	*Saïga tatarica*	saïga	X	
	Capreolus capreolus	chevreuil		*Capreolus capreolus*
	Sus scrofa	sanglier		***Sus scrofa***
4	*Mammuthus primigenius*	mammouth	X	
	Bos primigenius	aurochs		***Bos primigenius***
	Megaloceros giganteus	cerf géant	XX	
	Equus hydruntinus	cheval hydruntin	(XX)	(*Equus hydruntinus*)
absent	*Rhinocerotidae*	rhinocéros	XX	
	Ovibos moschatus	bœuf musqué	X	
	Ovis sp.	mouflon	X	
	Hemitragus bonali	tahr	X	
	Dama dama	daim		?
	Alces alces	élan		*Alces alces* ?
	Lepus timidus	lièvre		*Lepus* cf.*europaeus*
	Oryctolagus cuniculus	lapin		*Oryctolagus cuniculus*
absent	*Castor fiber*	castor		*Castor fiber*
	Marmota marmota	marmotte		(*Marmota marmota*)
1	***Canis lupus***	loup		***Canis lupus***
	Vulpes vulpes	renard roux		***Vulpes vulpes***
	Ursus arctos	ours brun		*Ursus arctos*
	Ursus spelaeus	ours des cavernes	X	
2	*Meles meles*	blaireau		*Meles meles*
	Alopex lagopus	renard polaire	X	
3	*Crocuta c.spelaea*	hyène	X	
	Lynx spelaea	lynx		*Lynx* sp.
	Felis silvestris	chat sauvage		*Felis silvestris*
	Martes martes/putorius	martre/putois		*Martes martes*
absent	*Panthera (Leo) spelaea*	lion	X	
	Panthera pardus	panthère	X	
	Cuon alpinus	dhôle	X	
	Gulo gulo	glouton	X	
	Lutra lutra	loutre		*Lutra lutra*
	Mustelidae	hermine, belette,etc.		?
Corse	*Megaloceros cazioti*	cerf de Caziot	XX	
	Cynotherium sardous	cuon sarde	XX	
	Prolagus sardus	lapin-rat		*Prolagus sardus*

(support de gravures), ne signifie pas pour autant l'existence de cette espèce dans l'environnement. Le cerf géant ne semble présent qu'à l'est avec une mention à Solutré et une au Pignon. On note l'absence totale d'éléments traditionnels froids tels que le rhinocéros laineux, le bœuf musqué ou le mouflon (ce dernier reflétant d'abord une topographie) ainsi que de taxons plus tempérés comme le daim ou l'élan. Il demeure un problème de détermination des restes de grands bovidés et il est souvent fait mention de 'cf.' avant le nom de genre. Il semble cependant que le bison des steppes soit nettement plus fréquent que l'aurochs. Le lièvre est moyennement représenté (et pas toujours identifié au niveau spécifique) et le lapin de garenne est rare (1 occurrence en Midi-Pyrénées).

Il existe toutefois des différences régionales parfois marquées : absence quasi-permanente du renne et du bison en Provence durant le dernier glaciaire (mais présence de l'aurochs) ; saïga particulièrement abondante en Gironde et en Aquitaine ; cerf et chevreuil présents en bordure ouest du Massif Central. L'existence d'associations comportant des «éléments tempérés» et des espèces plus «froides» (ex. renne et cerf se retrouvent associés dans 12 sites) semble prévaloir dans les faunes du dernier maximum glaciaire, avec quelques différences liées à des environnements contrastés et à la présence de zones plus clémentes (zone refuge), telles que la région Midi-Pyrénées ou la Provence.

Parmi les carnassiers, le loup, le renard et les ours (2 mentions d'ours bruns et 3 d'ours des cavernes) sont les plus fréquents. L'aire de répartition des ours des cavernes (*Ursus spelaeus*) semble se morceler durant cette période, ces populations relictuelles sont alors en voie de disparition, ce bien avant la fin du dernier glaciaire. Il en est probablement de même avec l'hyène des cavernes (1 occurrence) ; le lynx, le chat sylvestre et le putois restent particulièrement rares, mais probablement pour d'autres raisons. Par contre, le blaireau et le renard polaire sont plus régulièrement attestés. Parmi les carnivores manquants, il faut signaler le lion des cavernes (mais il est peint à la grotte Cosquer), la panthère, le dhole ou cuon (canidé encore présent en Asie) ainsi que les Mustélidés (glouton, hermine, belette,...).

En conclusion, les populations de grands mammifères au dernier maximum glaciaire en France correspondent à des écosystèmes riches et diversifiés. Une faune de type arctique (renne, cheval, bison) est caractéristique de cette période mais la situation géographique de la France a permis la permanence d'associations originales, en particulier dans le sud, avec des espèces plus tempérées. Malgré les rigueurs climatiques (présence de permafrost développé) on constate la présence de taxons (ex. sanglier) dans certaines zones (Massif central et bordure) et certains autres ont pu subsister dans des environnements topographiquement contrastés (vallées du Lot, Aveyron, Tarn). Il semble que le climat joue principalement sur les aires de distribution des espèces, certaines se développant au détriment d'autres qui verront leurs distributions se réduire (plus au sud, en Italie ou Ibérie), sans réellement connaître de disparition.

OPTIMUM HOLOCENE (8000 ± 1000 ans BP)

L'environnement physique (Morzadec-Kerfoun, Van Vliet-Lanoë, Antoine in Petit-Maire, 1999) est caractérisé par une transgression marine (environ -25 m) et une perte de l'englacement sur une grande partie du territoire. Si Les spectres polliniques (de Beaulieu in Petit-Maire, 1999) de la fin du Boréal sont globalement dominés par le noisetier associé à la chênaie mixte, il est cependant possible de mettre en évidence des provinces bien individualisées (biomes établis en fonction de l'importance de certains éléments de la chênaie).

Les données sur la grande faune du Tardiglaciaire de l'Holocène ancien sont nombreuses mais restent dispersées dans la littérature malgré quelques tentatives de synthèse continentales (Andersen *et al.*, 1990 ; Bridault et Chaix, ss-pr.) ou régionales (Bridault, 1993,1994, 1997b ; Chaix et Bridault, 1994 ; Limondin et *al.*, ss-pr.). Avec le développement de l'archéologie préventive, nombre d'entre elles ne figurent que dans des rapports d'analyse inédits et difficiles d'accès. C'est pourquoi, durant les années 90, dans le cadre du programme «*Dynamique de la Biodiversité et Environnement*», la communauté scientifique nationale concernée conjugue ses efforts pour réunir et valider ces informations dans une base de données cartographique informatisée. C'est à partir de cette base de données collective que sont issues les informations pour la tranche de temps de 8000 ± 1000 ans BP. Il résulte que les données présentées ici sont sans doute les plus complètes réunies à ce jour, mais qu'elles ne sont ni exhaustives ni toutes validées en matières de datation et de détermination taxinomique.

Nous avons opté pour une présentation la plus objective possible, sans extrapolation ni sélection des données (à une exception près, concernant le renne, cf. *infra*). Ainsi, le bouquetin, l'aurochs, le sanglier, etc., n'apparaissent pas dans le secteur sud-ouest en raison de l'absence de mentions les concernant dans la base de données préliminaire, mais on peut raisonnablement avancer que cela ne résulte que d'une lacune d'information, puisque toutes ces espèces figurent dans cette même région pour les tranches de temps immédiatement antérieure ou postérieure. Les données ont été réunies en 9 grands secteurs géométriques partageant la France continentale de manière à peu près équitable, la Corse constituant un dixième territoire.

Les données fauniques pour tranche de temps 8000 ± 1000 ans BP sont principalement issues de sites archéologiques du Mésolithique (Bintz in Petit-Maire, 1999), période qui est longtemps restée l'une des plus mal documentées par l'archéozoologie. Seuls 15 de nos 96 départements ont livré des informations concernant les grands mammifères (fig. 2). Ils se répartissent en 6 régions :

* les marges occidentales du Massif Central (Lot, Corrèze, Vienne),

* le Bassin Parisien (Eure, Somme, Seine-et-Marne),

* l'extrême Est (Moselle, Haut-Rhin, Haute-Saône, Doubs),

* le domaine alpin septentrional (Haute-Savoie, Isère),

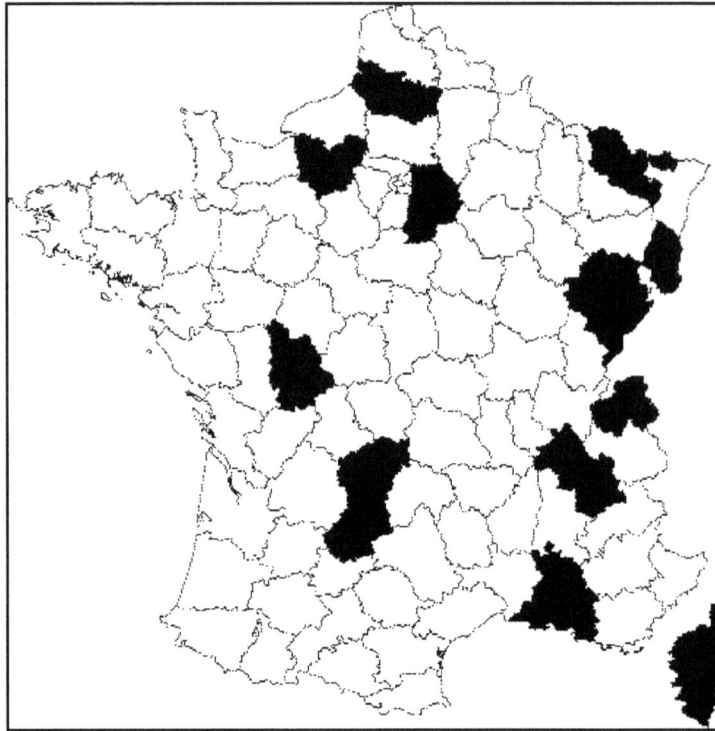

Figure 2. Répartition par département des mentions de grande faune à 8000 ± 1000 ans BP de l'*Atlas disachronique des mammifères de France* (état 1997, non complété, non validé).

* la Provence occidentale (Vaucluse, Bouches-du-Rhône),

* la Corse (Haute-Corse, Corse-du-Sud).

Les départements qui n'ont livré qu'un assemblage de grande faune de cette période ne sont pas rares. Une large bande atlantique, une grande partie du Sud-Ouest ainsi que le Centre, le Languedoc-Roussillon et son arrière pays sont vides de données fauniques. Cette répartition géographique reflète en partie celle des fouilles et des études archéozoologiques et en partie les conditions de préservation des vestiges osseux.

De plus, il faut préciser que rares sont les départements qui ont livré plus d'un assemblage de grande faune renvoyant à cette période, et que certains d'entre eux (Eure, Moselle, Haute-Saône) présentent une diversité taxinomique manifestement amoindrie par une taphonomie défavorable. Cette répartition géographique reflète également celle des fouilles et des études archéozoologiques. Certaines régions ont été jusqu'à présent négligées par la recherche en dépit de récents efforts ; c'est notamment le cas des zones d'altitude, particulièrement importantes pour cette époque en raison de la recolonisation forestière et de l'exploitation des certains de ces espaces par les chasseurs mésolithiques.

Principales caractéristiques de la grande faune mammalienne continentale (tabl. 1)

La période considérée correspond au Boréal et au début de l'Atlantique. Sur l'essentiel du territoire français continental, la grande faune froide du dernier Pléniglaciaire a fini de laisser la place au cortège tempéré, dominé par le cerf et le

sanglier. La recomposition des cortèges fauniques s'est en effet produite durant le Tardiglaciaire selon des modalités qui restent à préciser selon les régions.

Certaines espèces ont définitivement quitté la surface du globe dans la première moitié du Tardiglaciaire : *Ursus spelaeus, Crocuta spelaea, Panthera spelaea, Coelodonta antiquitatis, Mammuthus primigenius, Megaloceros giganteus.*

D'autres, comme l'antilope saïga, le renard polaire et peut-être aussi le glouton ont émigré alors vers des contrées septentrionales ou orientales bien avant 8000 ans BP. Le renne, gibier très fréquemment chassé, disparaît soudainement de la documentation hexagonale peu avant 12000 ans BP (Bridault et al., 2000). Son émigration semble être définitive, contrairement aux contrées septentrionales limitrophes (Belgique, Rhénanie du Nord) où sa présence saisonnière durant la dernière oscillation froide avant l'Holocène suggère un épisode d'extension méridionale de son aire de répartition. Par conséquent, l'unique mention du renne au Boréal (Vienne) semble irrecevable et n'a pas été considérée ici.

Dans le Bassin parisien comme dans l'Est de la France (Bridault et Chaix, 1999) par exemple, les données récentes laissent percevoir de forts contrastes entre les spectres fauniques d'une même région à la fin du Tardiglaciaire témoignant d'une recomposition rapide des spectres entre 12000 et 10000 ans BP. C'est alors que le cheval et le cerf sont particulièrement abondants dans les faunes chassées du Bassin parisien (Limondin et al., ss-pr. ; Bridault et al., 2001) et que le cheval disparaît quasiment de la documentation de cette région vers 10000 ans BP (Bridault, 1997a), pour réapparaître au cours du Néolithique. Il est toutefois encore attesté durant la période 9000-7000 ans BP dans la Vienne, la Haute-Saône, le Doubs et les Bouches-du-Rhône. Il continuera d'apparaître jusque dans les faunes de l'Atlantique récent, fréquemment dans le quart sud-ouest de la France, ailleurs plus sporadiquement.

Certaines espèces plus rarement attestées que les précédentes dans les spectres de chasse, n'ont pas pour autant totalement disparu des paysages de l'Holocène. Elles ont en effet pu s'adapter malgré des conditions moins propices, au moins dans certaines régions, et perdurer avec des densités peut-être plus faibles qu'auparavant et sur des territoires peu fréquentés par les chasseurs. C'est peut-être le cas de l'élan (*Alces alces*), qui déjà rarement attesté en France au Tardiglaciaire, s'est encore raréfié dans les faunes mésolithiques d'Europe de l'ouest, alors qu'il est abondant des contrées plus septentrionales à la même époque (Bridault, 1992). En France il est attesté vers 9000 ans BP dans la Somme et au cours du Boréal dans un site du Jura (seule région où il est fréquemment attesté au Tardiglaciaire). C'est dans cette même région que l'élan encore documenté au Néolithique et plus tardivement encore (Chaix et Desse, 1981). L'«âne» sauvage (*Equus hydruntinus*), également peu fréquent, était encore présent vers 8000 ans BP dans les Bouches-du-Rhône au moins.

C'est également le cas du bison (*Bison bonasus*) et de la marmotte, très rares dans les spectres mésolithiques du Boréal. Bien que non mentionnés, ils sont attestés pour des périodes ultérieures de l'Holocène respectivement dans l'Est et le Jura, et dans les Alpes. C'est aussi le cas d'espèces comme le bouquetin, le chamois et le lièvre variable (*Lepus timidus*, ici non différencié de *L. europaeus*), uniquement mentionnées dans les Alpes septentrionales et sans doute déjà réfugiées en altitude, comme dans les Pyrénées et peut-être aussi dans le Massif Central (Bridault et Fontana, 2001). La raréfaction des espèces aujourd'hui considérées comme alpines, constatée à partir du début du Mésolithique, traduit probablement la réorganisation verticale des aires de répartition de ces espèces, remontées en altitude à la faveur de la rapide reconquête forestière (Bridault et Chaix, 1999). Les rares spectres fauniques d'altitude connus attestent seulement la chasse d'ongulés (bouquetin et chamois).

Les carnivores présents au Tardiglaciaires sont en revanche encore bien représentés entre 9000 et 7000 ans BP dans toutes les régions documentées. Le loup et le renard sont mentionnés dans les zones non montagnardes, du Bassin Parisien au Quercy et en Provence. L'ours brun et le lynx (*Lynx* sp.) figurent en montagne comme dans les zones de basse altitude qu'ils ne déserteront qu'au Sub-Atlantique. Le chat sauvage, le blaireau et la martre sont attestés dans toutes les régions, et la loutre est abondante près des cours d'eau du Bassin parisien au moins. Elle y côtoie le castor, mentionné par ailleurs dans toutes les autres régions continentales documentées. Le lièvre (*L.*cf. *europaeus*) est également présent en dépit du fort boisement. Le lapin en revanche, reste cantonné aux régions méditerranéennes et au pourtour de l'Aquitaine, dont il ne sortira qu'au Moyen Âge central, transféré par l'homme jusqu'en Écosse et en Scandinavie.

Le cerf et le sanglier sont attestés dans les 13 départements de France continentale documentés pour la période 9000-7000 ans BP. Dans 11 d'entre eux, ils sont accompagnés du chevreuil et de l'aurochs. Dans les plaines comme dans les montagnes, ces quatre espèces formaient le fonds du peuplement de grands mammifères et constituaient le principal gibier des Mésolithiques (Bridault, 1997a,b).

De cette description rapide, on pourrait retirer l'impression qu'aux environs de 8000 ans BP, la faune actuelle est bien en place et qu'elle ne varierait plus guère jusqu'à nos jours si ce n'est en régressant sous la pression anthropique. Ce serait négliger la présence, au Mésolithique, dans le Doubs, les Bouches-du-Rhône et la Somme, de restes de chien domestique (*Canis familiaris*) et, surtout l'apparition en Corse et en quelques points du littoral méditerranéen, peu après 6600 ans BP (entre 5600 et 5000 cal. av. J.C. selon les régions), des premiers moutons (*Ovis aries*), chèvres (*Capra hircus*), vaches (*Bos taurus*) et porcs (*Sus s. domesticus*) domestiques, importés de proche en proche depuis l'Orient, avec l'avancée de la néolithisation. Ce mouvement, renforcé peu après par l'afflux de populations domestiques des mêmes espèces acheminées par le courant danubien jusque dans le Nord et l'Est du Bassin Parisien, aboutira à un bouleversement profond de la grande faune, les espèces domestiques entrant partout en concurrence avec les ongulés sauvages, modifiant profondément leurs comportements écologiques et finissant par les supplanter sur une importante partie du territoire. Nul doute qu'une comparaison entre les cartes 8000 et 1000 ans BP ferait ressortir autant de différences qu'entre les cartes 18 000 et 8000 ans BP.

La situation insulaire corse

Entre 9000 et 7000 ans BP, la faune mammalienne de Corse, pour laquelle on dispose d'une documentation assez abondante, différait totalement de celle du continent. Aucune des espèces attestées sur ce dernier n'y était présente. Aux côtés de trois taxons endémiques de micromammifères, on n'y trouvait qu'un petit lagomorphe, lui-même endémique, *Prolagus sardus*. Cette situation résulte de deux phénomènes :

- un isolement poussé, depuis le Pléistocène moyen au moins ; les quelques grandes espèces qui y ont résisté ont évolué en milieu clos vers des formes propres au massif insulaire corso-sarde, dont il ne subsistait, au début du Tardiglaciaire, qu'un cervidé (*Megaloceros cazioti*), un canidé (*Cynotherium sardous*) et *Prolagus sardus* ;

- l'extinction du cervidé et du canidé, à la fin du Tardiglacaire (Vigne, 1999). Ont-ils succombé au réchauffement tardiglaciaire ou sont-ils tombés sous les flèches de chasseurs épipaléolithiques inconnus à ce jour ? Pour l'instant, les plus anciennes attestations de la présence de l'homme moderne sur l'île sont datées entre 9000 et 8000 ans BP, et aucun des neuf sites archéologiques corso-sardes qui documentent cette période n'a livré le moindre reste de ces deux espèces (Vigne, 2000).

Ainsi, autour de 8000 ans BP, la diversité de la faune mammalienne de Corse a connu son minimum absolu, compensé, à partir du début du Néolithique, par l'introduction anthropique des espèces d'origine continentale qui constituent le peuplement diversifié actuel de l'île (Vigne, 1999).

CONCLUSIONS

La composition des grands mammifères au LGM représente un mélange d'espèces de toundra ou de forêt sub-arctique (renne, renard polaire) et de steppes arborées froides (cheval, bison, mammouth), de prairies et de forêts caducifoliés (aurochs, cerf, mégacéros, chevreuil, lynx,...), de montagne (bouquetin chamois). Parmi les herbivores, on note l'absence totale d'éléments traditionnels froids tels que le rhinocéros laineux, le bœuf musqué ou le mouflon, ainsi que de taxons plus tempérés comme le daim ou l'élan ; les carnivores sont encore abondants. La France, avec des reliefs développés au Sud, montre des cloisonnements qui conditionnent une répartition en mosaïque des communautés de grands mammifères avec conservation au sud d'espèces les plus sensibles au froid, alors que le nord semble plus homogène avec une relative moindre diversité et une association de type arctique.

A l'Holocène ancien, la situation contraste par une plus grande homogénéité dans la composition des associations fauniques : principalement cerf, sanglier, chevreuil, et dans une moindre mesure, aurochs. L'essentiel du territoire français continental montre cette association typique de conditions tempérées, avec en montagne, une spécificité des spectres d'altitude (qui comptent aussi du bouquetin et du chamois). Les cortèges de carnivores enregistrent également sensiblement ce retour des grandes forêts de feuillus : martre, chat sauvage sont récurrents dans les sites. L'arrivée des espèces domestiques qui s'ajoute à ce fond commun, amorce un nouveau bouleversement majeur dont l'homme est cet fois le principal protagoniste

Entre ces deux extrêmes climatiques, on voit la disparition de notre territoire de nombreuses espèces de mammifères (tabl.1). Celles-ci ne semblent pas disparaître selon les mêmes rythmes : si, par exemple, la plupart des grands prédateurs (ours des cavernes, hyènes, panthère, glouton, dhole) ne dépassent que rarement la limite LGM[4], certains herbivores subsistent au moins jusqu'aux environs de 12 000 ans BP, tel que le renne en particulier (Bridault et Fontana, 2001) et parfois perdurent de façon ponctuelle comme le cheval (Brugal *et al*, 2001). De fait, la plupart des espèces changent leurs aires de distribution au sein de l'Europe de l'ouest et au-delà, ou disparaîtront plus tard (mammouth, mégacéros, ours des cavernes). Il n'existe donc pas de réelle crise biologique contemporaine de la fin du dernier glaciaire avec extinction massive d'espèces. Par ailleurs, à l'Holocène on constate que sont apparus la loutre, le castor et l'élan (ponctuel) ainsi que du premier animal domestique : le chien. Malgré ces rajouts, la faune " glaciaire " désigne des associations plus riches (plus de trente espèces) que celle du Boréal et du début de l'Atlantique (près de 22 espèces). Le

détail de ces ajustements biologiques reste à être précisé et ils sont vraisemblablement à mettre en relation avec des processus climatiques complexes, souvent de haute résolution (*i.e.*, événements d'Heinrich avec le H1 daté d'env. 14,5 ka ou le H0 - 'younger Dryas' - daté d'env. 12 ka). Cette dernière limite est probablement plus pertinente que celle proposée classiquement à 10 ka pour le début de la période Holocène.

La diversité des associations traduit la complexité des processus climatiques qui façonnent l'histoire des écosystèmes et vont mener par la réduction des aires de distribution ou la disparition de certaines espèces, au panorama de la faune sauvage du début de l'Holocène. La succession des associations de mammifères retrace non seulement l'évolution des changements conduisant à l'installation de la faune actuelle mais nous renseigne également sur l'histoire des liens que l'homme préhistorique a entretenus avec les animaux sauvages. La disparition plus ou moins progressive (morcellement des aires) des grands prédateurs (ours, hyène, lion), le remplacement d'ongulés par des taxons présentant d'autres caractéristiques éco-éthologiques (cerf *vs* renne, aurochs *vs* bison) vont impliquer des stratégies alimentaires et sociales spécifiques.

Adresses des auteurs

Jean-Philip BRUGAL
UMR 6636 – MMSH – BP647
13094 Aix-en-Provence, FRANCE
brugal@mmsh.univ-aix.fr

Anne BRIDAULT
UMR 7041- Maison de l'Archéologie et de l'Ethnologie
21 allée de l'Université
92023 Nanterre cedex France
bridault@mae.u-paris10.fr

Jean-Luc GUADELLI
UMR 9933 - Univers.Bordeaux I
av. des facultés
33405 Talence cedex France
guadelli@iquat.u-bordeaux.fr

Jean-Denis VIGNE
ESA 8045, MNHN, lab.Anatomie comparée
55 rue Buffon
75005 Paris France
vigne@cimrs1.mnhn.fr

Bibliographie

Seuls sont indiqués ici quelques ouvrages généraux complétés pour exemples de quelques articles plus ponctuels. Un certain nombre de données sur les gisements de cette période reste encore inédit. Les dates 14C des gisements se trouvent dans la revue Radiocarbone.

ALLARD M., 1995 – Magdalénien ancien (=Badegoulien) et Magdalénien moyen aux Peyrugues (Lot). *Annales Rencontres Archéologiques de Saint-Céré*, 4 : 1-13

ANDERSEN S.H. et alii, 1990.- Making cultural ecology relevant to Mesolithic research: I. A data base of 413 Mesolithic faunal

---

[4]   On relèvera la présence du Lion dans le site azilien du Closeau (Hauts-de-Seine) entre 12850 et 12191 cal. BC, dernier témoin de cette espèce en Europe du Nord (Bémilli 2000).

assemblages. *In*: P.M. Vermeersch & Ph. Van Peer (éds.), *Contributions to the Mesolithic in Europe*. Leuven : Leuven University Press, pp. 23-51.

ANTOINE P., 1991 - Nouvelles données sur la stratigraphie du Pléistocène supérieur de la France septentrionale, d'après les sondages effectués sur le tracé du TGV Nord. Publ. Du Centre d'Etudes et de Recherches Préhistoriques, Univers. de Lille, 3, pp.9-20

BARD E., HAMELIN B., FAIRBANKS R.G., ZINDLER A., 1990.- Calibration of the ^{14}C time scale over the last 30,000 years using mass spectrometric U-Th ages from Barbados corals. *Nature, 345* :405-410.

BAZILE F., 1990- Le solutréen et épisolutréen du Sud-Est de la France. *In Les industries à pointes foliacées du Paléolithique supérieur européen*, Krakow 1989, Liège : *ERAUL*, 42 : 393-423

BEMILLI C., 2000 - Nouvelles données sur les faunes aziliennes du Closeau, Rueil-Malmaison (Hauts-de-Seine), *Actes de la Table Ronde de Chambéry*, 12-13 mars 1999, *Le Paléolithique supérieur récent : nouvelles données sur le peuplement et l'environnement*, G.Pion (ss.la dir.), S.P.F., Mém. 28 : 29-38

BRIDAULT A., 1992.- The status of Elk during the Mesolithic. *Anthropozoologica*, 16 :151-159.

BRIDAULT A., 1994.- Les économies de chasse épipaléolithiques et mésolithiques du Nord et de l'Est de la France : nouvelles analyses. *Anthropozoologica,* 19 : 55-67.

BRIDAULT A., 1997a.- Chasseurs, ressources animales et milieux dans le Nord de la France de la fin du Paléolithique à la fin du Mésolithique : problématique et état de la recherche. In : *119e Congr. nat. soc. hist. scient.* (Amiens, 1994), *Pré- et Protohistoire*. Paris : CTHS,, p. 165-176.

BRIDAULT A., 1997b.- Broadening and diversification of hunted resources, from the Late Palaeolithic to the Late Mesolithic, in the North and East of France and the bordering areas. *Anthropozoologica*, 25-26 : 295-308.

BRIDAULT A., CHAIX L., 1999.- Contribution de l'archéozoologie à la caractérisation des modalités d'occupation des sites alpins et jurassiens, de l'Epipaléolithique au Néolithique ancien In L'Europe des derniers chasseurs : Epipaléolithique et Mésolithique, Actes 5é Congrès Int., Paris : CTHS, p.547-558

BRIDAULT A., CHAIX L., ss-pr..- Ruptures et équilibres dans la grande faune à la fin du Pléistocène et durant l'Holocène ancien en Europe occidentale. In : *Equilibres et ruptures dans les écosystèmes durant les 20 derniers millénaires en Europe de l'Ouest, Actes du colloque international de Besançon*, septembre 2000. Besançon : Presses Universitaires Franc-Comtoises.

BRIDAULT A., CHAIX L., PION G., OBERLIN, C., THIEBAULT, S. et ARGANT, J. 2000.- Position chronologique du Renne (*Rangifer tarandus* L.) à la fin du Tardiglaciaire dans les Alpes du Nord françaises et le Jura méridional. *In G. Pion ed., Actes de la table ronde de Chambéry, Le Paléolithique supérieur récent : nouvelles données sur le peuplement et l'environnement*. Société Préhistorique Française, mémoire XXVIII, Paris, p.47-57..

BRIDAULT A., FONTANA L, 2001 - Enregistrement des variations environnementales par les faunes chassées dans les zones de moyenne montagne d'Europe occidentale, au Tardiglaciaire et au début de l'Holocène. XIVe Congrés UISPP, Liège, 2-8 Sept.2001, vol.pré-actes : 64

BRUGAL J.-P., 1981 - Les faunes de grands mammifères du Pléistocène terminal en Languedoc oriental. *Et.Quatern. Languedoc.*, n°spécial A.F.E.Q. Mai 1981: 21-28

BRUGAL J.P., BUISSON-CATIL J., HELMER D., 2001 - L'aven des Fourches II (Sault, Vaucluse) : les derniers chevaux sauvages en Provence. *Paléo*, 13 : 73-88

BRULHET J., PETIT-MAIRE N. (Dir.), 1999 - *La France pendant les deux derniers extrêmes climatiques. Variabilité naturelle des environnements*, Châtenay-Malabry, ANDRA / CNF-INQUA, 67 p. (2 cartes au 1 : 1.000.000)

CHAIX L. et BRIDAULT A., 1994 (1992).- Nouvelles données sur l'exploitation des animaux sauvages de l'Epipaléolithique au Mesolithique final dans les Alpes du nord et le Jura. *Preistoria Alpina* : 115-127.

CHAIX L., et DESSE J., 1981.- Contribution à la connaissance de l'élan (*Alces alces* L.) postglaciaire du Jura et du Plateau suisse. Corpus de mesures. *Sonderdruck aus Quartär* 31/32:139-190.

CLOTTES J., COURTIN J., CLLINA-GIRARD J., ARNOLD M., VALLADAS H. – 1997 – News from Cosquer cave : climatic studies, recording, sampling, dates. *Antiquity*, 71(272) : 321-326

CNRS (éd.) – 1979 – *La fin des temps glaciaires en Europe*. Coll.Intern n°271, 2 vol., 892p.

DELPECH F., 1983 – Les faunes du Paléolithique supérieur dans le Sud-Ouest de la France. *Cahier du Quaternaire*, 6, éd.C.N.R.S., 453 p.

D'ERRICO F., 1994 - Birds of the Cosquer cave. The great auk (*Pinguinus impennis*) and its significance during the Upper Palaeolithic. *Rock Art Research*, 11 (1), pp.45-57

FAGNART J.P., 1988 - Les industries lithiques du Paléolithique supérieur dans le Nord de la France. *Rev.Archaéol.de Picardie*, n°sp., 153 p.

GRIGGO C. (ss.pr.) – La faune de la grotte du Placard. Etudes paléontologique, paléoenvironnementale et archéozoologique. In *La Grotte du Placard*, J.Clottes, L.Duport, V.Feruglio (éds.)

HUNTLEY B., BIRKS H.J.B., 1981.- *An atlas of past and present pollen maps for europe : 0-13000 years ago*. Cambridge : Cambridge University Press.

LABEYRIE L., JOUZEL J., 1999 - Les soubresauts millénaires du climat. La Recherche, 321 : 60-61

LIMONDIN-LOZOUET N., BRIDAULT A., LEROYER C., PONEL P., ANTOINE P., CHAUSSE C., MUNAUT A.V., PASTRE J.F., ss-pr. - Evolution des écosystèmes de fond de vallée en France septentrionale au cours du Tardiglaciaire : l'apport des indicateurs biologiques. *Actes du Colloque «Paléohydrologie des 15 derniers millénaires» Motz Juin 2001*, J.P. Bravard et M. Magny (éds.)

PETIT-MAIRE N., Coordination Scientifique, avec la collaboration de, Antoine P., Beaulieu (de) J.-L., Bintz P., Brugal J.-P., Girard M., Morzadec M.-T., Renault-Miskovsky J., Roblin-Jouve A., Van Vliet-Lanoë B., 1999 - *La France à l'optimum climatique holocène : 8 000 ± 1 000 ans B.P.*, Paris, ANDRA / CNF-INQUA / IGN (1 carte au 1 : 1 000 000).

PETIT-MAIRE N., Coordination Scientifique, avec la collaboration de, Antoine P., Beaulieu (de) J.-L., Bintz P., Brugal J.-P., Girard M., Morzadec M.-T., Renault-Miskovsky J., Roblin-Jouve A., Van Vliet-Lanoë B., 1999 - *La France au dernier maximum glaciaire : 18 000 ± 2 000 ans B.P.*, Paris, ANDRA / CNF-INQUA / IGN (1 carte au 1 : 1 000 000).

ROZOY J.-G., 1978.- *Les derniers chasseurs. Bull. archéol. champenoise*, n°spécial, 3 vol., 1256 p.

SALOTTI, M., BELLOT-GOURLET, L., COURTOIS, J.-Y., DUBOIS, J.-N., LOUCHART, A., MOURER-CHAUVIRÉ, C., OBERLIN, C., PEREIRA, E., POUPEAU, G. ET TRAMONI, P., 2000.- La fin du Pléistocène supérieur et le début de l'Holocène en Corse : apports paléontologique et archéologique du site de Castiglione (Oletta, Haute-Corse). *Quaternaire*, 11, 3-4 : 219-230.

SERONIE-VIVIEN M.R., 1995 – La grotte de Pégourié, Caniac-du-Causse (Lot). *Préh.Querc.* suppl. n°2, 392p.

SOFFER O., GAMBLE C. (eds.) – 1990 – *The World at 18.000 BP, Northern Latitudes*. London : Allen & Unwin, 351p.

VIGNE J.-D., 1999.- The large "true" Mediterranean islands as a model for the Holocene human impact on the European vertebrate fauna ? Recent data and new reflections. In : N. Benecke (éd.), *The Holocene history of the European vertebrate fauna. Modern aspects of research* (Workshop, 6th-9th April 1998, Berlin). Berlin : Deutsches Archäologisches Institut, Eurasien-Abteilung, p. 295-322 (*Archäologie in Eurasien*, 6).

VIGNE J.-D., 2000.- Les chasseurs préhistoriques dans les îles méditerranéennes. *Pour la Science*, Dossier Hors série " *La valse des espèces* ", Juillet 2000 : 132-137.

RÉCONCILIATION DES MODELES « ÉVOLUTION MULTIRÉGIONALE » ET « SORTIE D'AFRIQUE »...

Valéry ZEITOUN

Résumé : Plusieurs travaux montrent que les changements de faune et de flore sont les conséquences d'événements climatiques globaux autour de 2,5 Ma, 1,8 Ma et 1,0 Ma. Des données environnementales et anatomiques indépendantes sont confrontées pour tenter de comprendre quels sont les mécanismes évolutifs qui agissent au sein du genre Homo. Ainsi, le résultat phylogénétique fondé sur l'analyse cladistique de 35 unités taxinomiques et 468 caractères crâniens montre l'existence d'une radiation en Afrique de l'est au sein d'un grade *Homo habilis* avec au moins quatre espèces différentes. Cette radiation humaine est congruente avec celles, contemporaines, des Bovidés et des Cercopithecidés est-africains à partir de 2,5 Ma. Quand une autre pulsation climatique se produit vers 1,8 Ma, c'est l'opportunité pour certains *Homo erectus* ou *Homo ergaster* de quitter l'Afrique vers l'Eurasie. Autour de 1,0 Ma les premiers Homo sapiens apparaissent en Afrique de l'est, se répandent « rapidement » vers le nord et le sud et atteignent l'Eurasie. Ils acquièrent alors des « sous-spécificités » régionales. Les hommes modernes remplacent ensuite toutes les sous-espèces humaines et atteignent l'Australie vers 35 000 ans selon un schéma qui permet de réconcilier les deux modèles principaux de « sortie d'Afrique » et d'«évolution multirégionale ».

Abstract: Several works provide evidence of faunistic and floristic shifts as consequences of global climatic events around 2.5 Myrs, 1.8 Myrs and 1.0 Myrs. Independant environmental and anatomical data are confronted to try to understand the evolutionary process among the genus Homo. Thus, the result of a cladistic analysis based on 35 taxonomical units and 468 features of the calvaria is showing the onset of East African speciations among the grade of *Homo habilis* with at least four splitted species. This human radiation is congruent with those contemporaneous of Bovids and Cercopithecids in East-Africa. When another climatic pulse happens around 1.8 Myrs, it is a good opportunity for some *Homo erectus* and/or *Homo ergaster* to leave Africa to Western and Eastern Asia. Around 1.0 Myrs the first Homo sapiens appear in Africa and « quickly » spread northwards and southwards, also in Eurasia. Then, everywhere, they get regional « subspecificity ». Finally, modern humans, as the current unique survivor of the genus Homo, replace all other human subspecies and reach Australia around 35 000 years according to a scheme that reconciles « Out of Africa » and « multiregional » models.

INTRODUCTION

Depuis une dizaine d'années, un faisceau d'indices issus des recherches de terrain suggère à la fois une origine plus ancienne et une diversification plus précoce pour *Homo* que cela n'était envisagé précédemment (Howell, 1999). Il est montré ici quel est l'impact de la révision de la définition d'*Homo erectus* quant à l'origine et à l'expansion du genre *Homo*. Du fait du paradigme courant, principalement basé sur une définition chronologique de l'espèce, il n'y a pas de différence marquée parmi les espèces du genre *Homo*. L'attention est portée sur *Homo erectus* parce que c'est un taxon dont la définition a changé en fonction des différents paradigmes scientifiques. Aussi, c'est un point de vue basé uniquement sur l'anatomie qui est proposé ici.

Un travail réel de systématique se doit d'établir la taxinomie à partir des spécimens - objets réels – jusqu'à des niveaux supérieurs : populations, sous-espèces, espèces – entités qui demeurent théoriques en paléoanthropologie - avant de réaliser le travail de phylogénie. L'étude de la structure de parenté étant une condition préalable à celle des processus évolutifs, un moyen de résoudre le problème est de réaliser une analyse cladistique. Mais, il faut garder à l'esprit que les fossiles ne sont qu'un reflet partiel de la réalité biologique. Aussi est-il utile de prendre en compte toutes les données disponibles ; morphologiques et métriques. Suivant ce principe, une analyse détaillée de la *calvaria* -complexe anatomique le mieux préservé parmi les spécimens attribués à *Homo erectus* – a permis d'établir une matrice de 345 caractères métriques et 123 caractères morphologiques qui a été appliquée à 32 crânes d'Hominidés fossiles bien conservés et des échantillons modernes d'*Homo*, *Pan* et *Gorilla* (Zeitoun, 2000a). Le résultat de l'analyse cladistique est un arbre unique (figure 1).

Dans cet arbre il est possible d'associer les spécimens et de proposer un nom quand un type, un holotype ou un paratype est présent dans l'ensemble obtenu. Ainsi, et selon le principe hennigien, quatre espèces sont identifiables à la base du genre *Homo* puisque les quatre spécimens décrits ici sont de même niveau taxinomique. Or, deux d'entre eux ont une attribution spécifique : *Homo rudolfensis* Alexeev, 1986 pour Knmer 1470 dont il est le type et, *Homo ergaster* Groves et Mazak 1975 pour Knmer 1813 selon l'acception de Groves (1989, p. 239). Ce résultat, bien qu'original, s'accorde avec l'avis de Lieberman *et al.*, (1996) qui considèrent Knmer 1813 comme appartenant à une lignée pouvant être le groupe frère d'*Homo erectus* et celui de Stringer (1986 p.290) qui, discourant sur l'appartenance taxinomique de ces fossiles, écrit qu'une radiation peut avoir existé au début de la lignée humaine avec au moins trois espèces plio-pléistocènes de «early» *Homo*. Ici il y en a quatre !

Un clade qui regroupe les spécimens indonésiens de Trinil – spécimen type de l'espèce - et Sangiran et, le spécimen africain de Nariokotome peut être nommé *Homo erectus*. A l'étape suivante, l'information peut être interprétée de différentes manières. Et c'est bien d'interprétation dont il s'agit à ce niveau. La première solution prend en compte les espèces *Homo rhodesiensis* en Afrique (C) et *Homo soloensis* en Asie (D), puis *Homo sapiens* (E). La seconde utilise les

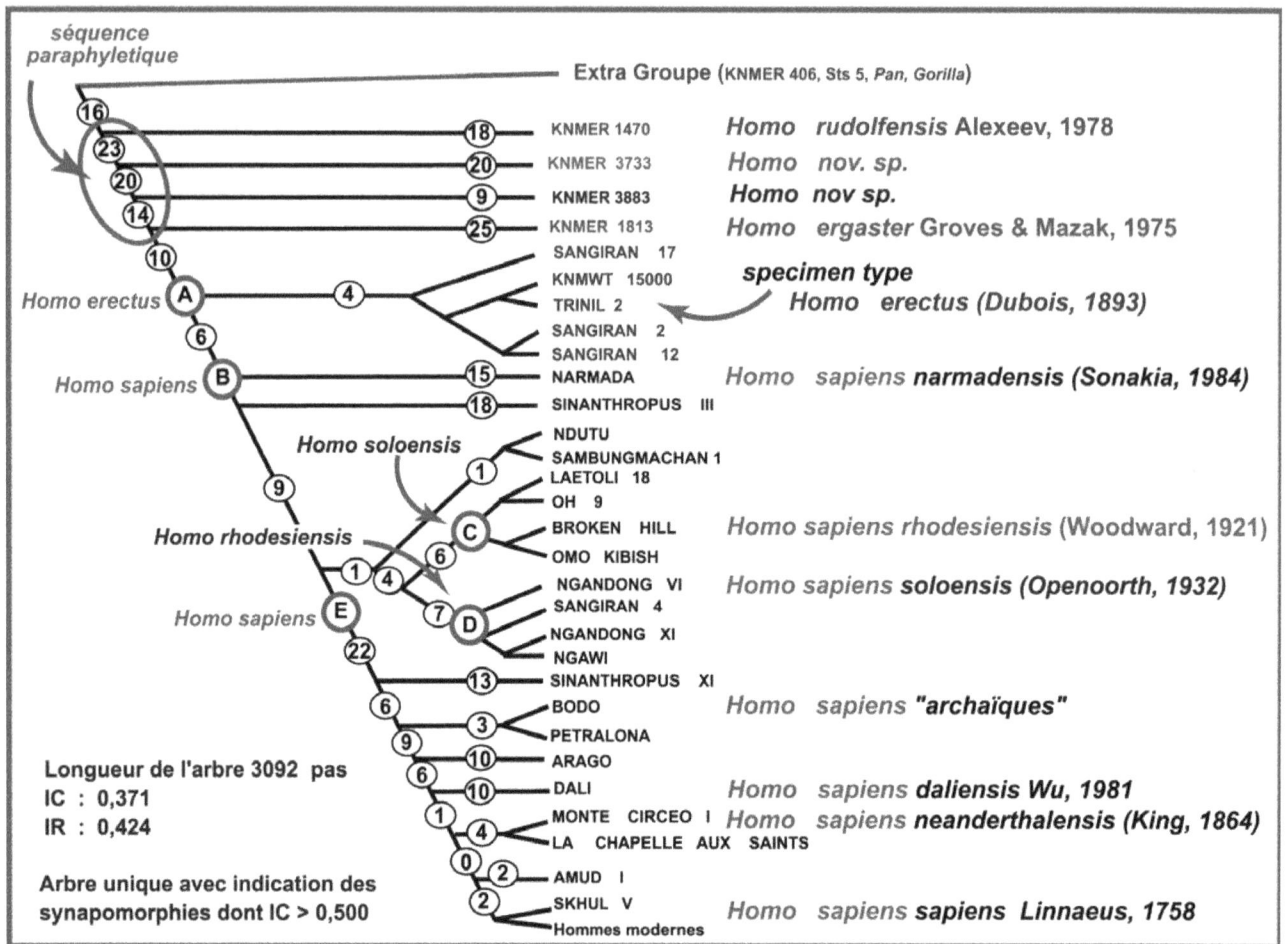

Figure 1 - Hypothèse phylogénétique incluant le nom des types, holotypes et paratypes

noms de sous-espèces d'*Homo sapiens* (A) pour décrire les unités taxinomiques : *Homo sapiens narmadensis* en Inde (Kennedy *et al.,* 1991), *Homo sapiens rhodesiensis* en Afrique (Campbell, 1972), *Homo sapiens soloensis* (Dubois, 1940 ; Tobias, 1985 ; Stringer, 1987 ; Brauer & Mbua, 1992) en Indonésie, diverses sous-espèces d'*Homo sapiens* dites « archaïques » avec *Homo sapiens pekinensis* (Zeitoun, 2000a) et *Homo sapiens daliensis* (Zhou *et al.,* 1982) en Chine et *Homo sapiens neanderthalensis* et *Homo sapiens sapiens*. Pour argumenter et tester cette hypothèse phylogénétique, il faut la confronter aux données environnementales et chronologiques pour voir si elle permet d'établir un scénario évolutif cohérent.

LES DONNÉES CHRONOLOGIQUES

Si le schéma phylogénétique proposé ici est placé dans un cadre chronologique, alors, cela suppose une origine très ancienne pour *Homo sapiens* (figure2). C'est justement ce que suggèrent les découvertes de terrain de ces dernières années. En effet, si les spécimens d'Atapuerca ont été décrits comme de possibles ancêtres communs aux hommes modernes et aux néandertaliens (Bermudez *et al.*, 1997) ici respectivement *Homo sapiens sapiens* et *Homo sapiens neanderthalensis*, alors ce peuvent être des *Homo sapiens*. Et, les plus anciens ont 0,8 Ma. C'est également le cas pour

le spécimen italien de Ceprano souvent rapporté à ceux d'Atapuerca et qui est daté à 0,9 Ma (Ascenzi *et al.*, 2000.).

Un autre indice confortant les conséquences de l'hypothèse posée ici provient du crâne érythréen de Danakil d'environ 1,0 Ma décrit par ses inventeurs (Abbate *et al.*, 1998) comme ayant déjà des caractères d'*Homo sapiens*. Si il y a ici un faisceau d'arguments supportant une origine ancienne pour *Homo sapiens*, il est également des données chronologiques autour de 40 000 ans pour Ngandong et Sambungmachan (Swisher *et al.*, 1996 ; Falgueres *et al.*, ce congrès.) qui peuvent conduire certains anthropologues à accepter plus facilement ces spécimens comme n'appartenant pas à *Homo erectus* conformément à l'hypothèse proposée ici (figure 3).

LES DONNÉES ENVIRONNEMENTALES ET CLIMATIQUES

Géologie et tectonique à 2,5 Ma

En ce qui concerne l'environnement, plusieurs événements globaux et géologiques conduisent à un changement climatique majeur vers 2,5 Ma avec la surrection des montagnes transantarctiques et andines, la clôture de l'isthme de Panama et un « rifting » en Afrique orientale, l'apparition de deux pôles glaciaires avec l'arrivée des eaux froides dans

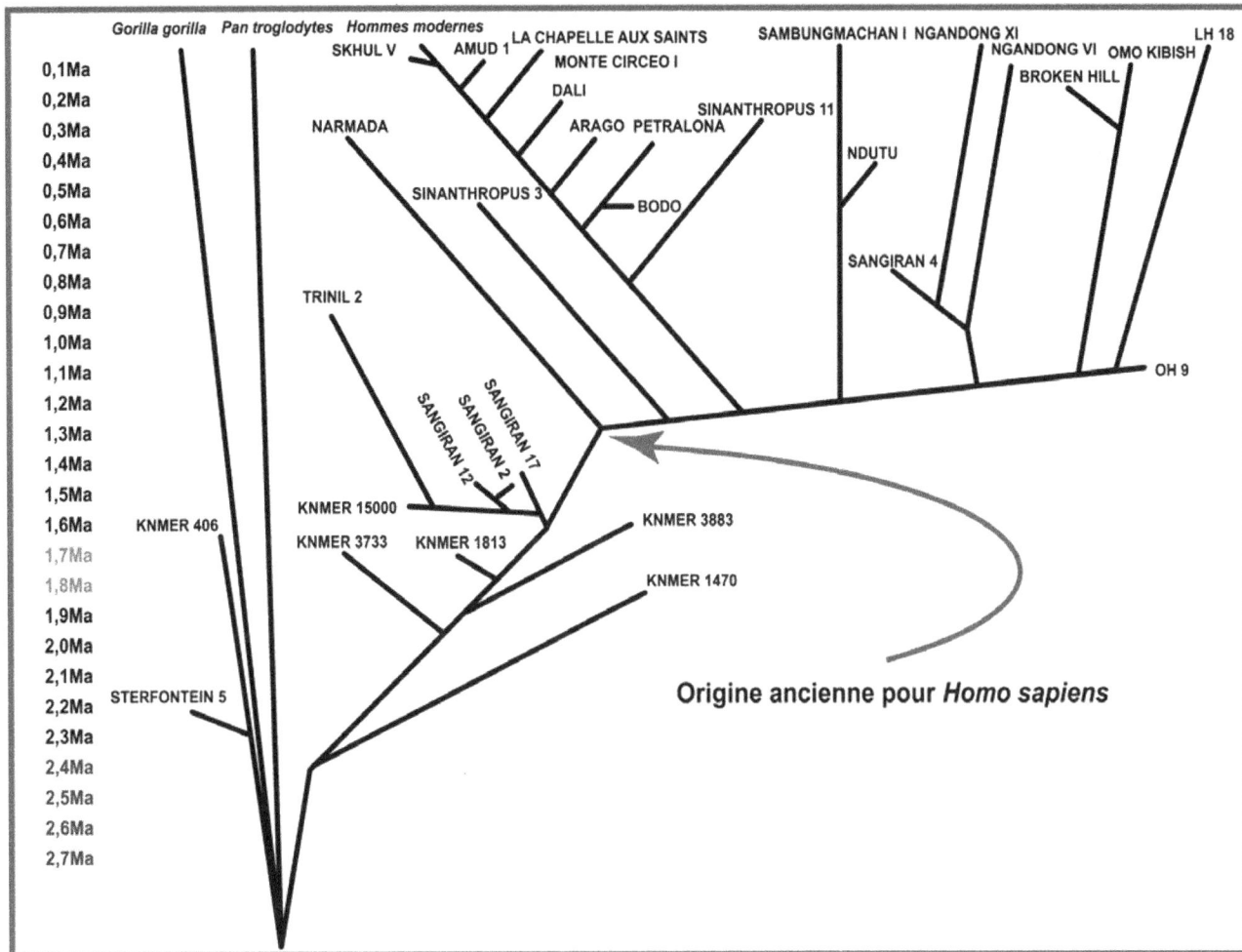

Figure 2 - Hypothèse phylogénétique dans le cadre chronologique

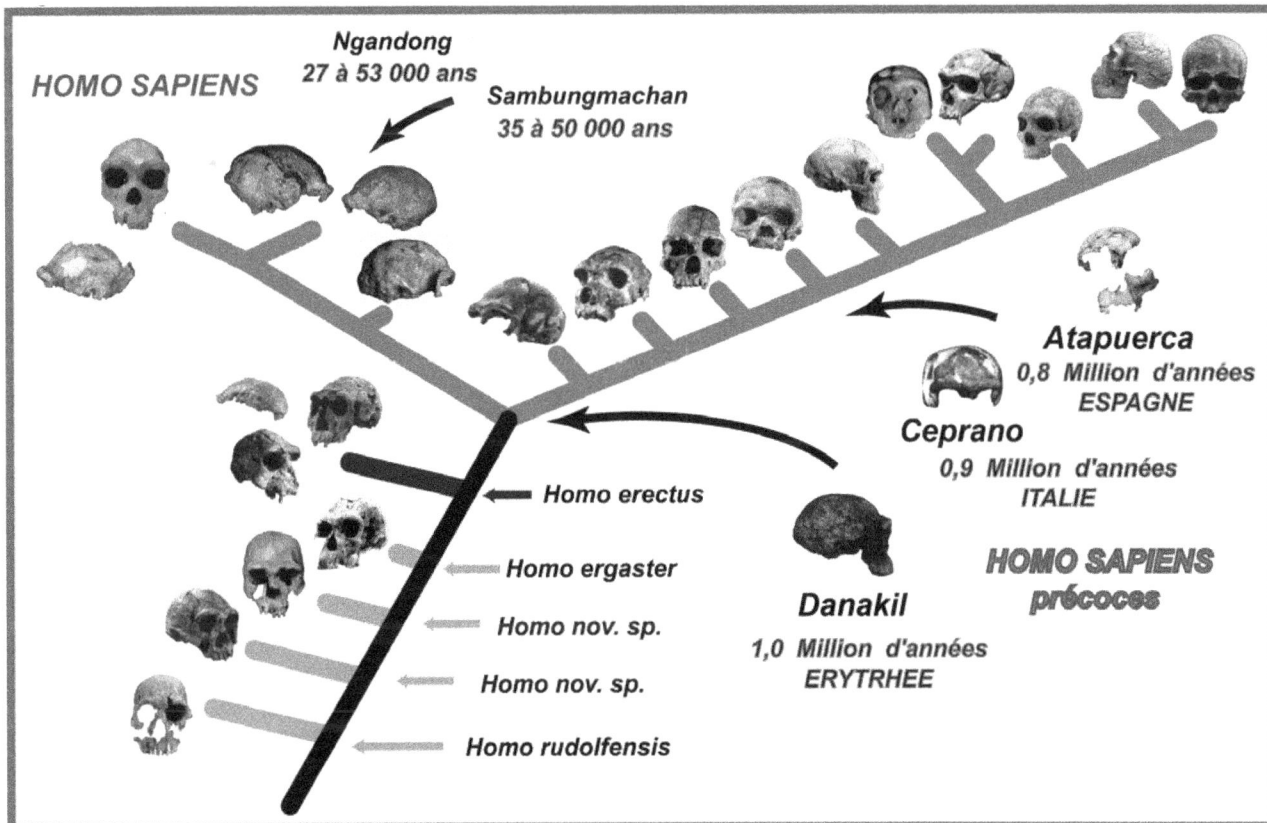

Figure 3 - Hypothèse phylogénétique et interprétation taxinomique

l'Atlantique sud, des dépôts loessiques en Alaska et en Chine puis, une régression marine (Golfe du Mexique et Méditerranée) et des dépôts éoliens au large des côtes africaines subsahariennes (figure 4).

La flore à 2,5 Ma

Les changements climatiques ont un impact sur la flore avec les derniers conifères en zone arctique et des taxons africains

qui révèlent des conditions plus contrastées conduisant à une ouverture du paysage (figure 5).

Les enregistrements palynologiques recueillis au large de la côte nord-atlantique de l'Afrique révèlent qu'une réduction progressive de la végétation de savane serait le résultat du développement d'un désert dans le Sahara occidental, enregistrements confortés par des données identiques au large de la Namibie vers 2,5 Ma. Les événements polliniques

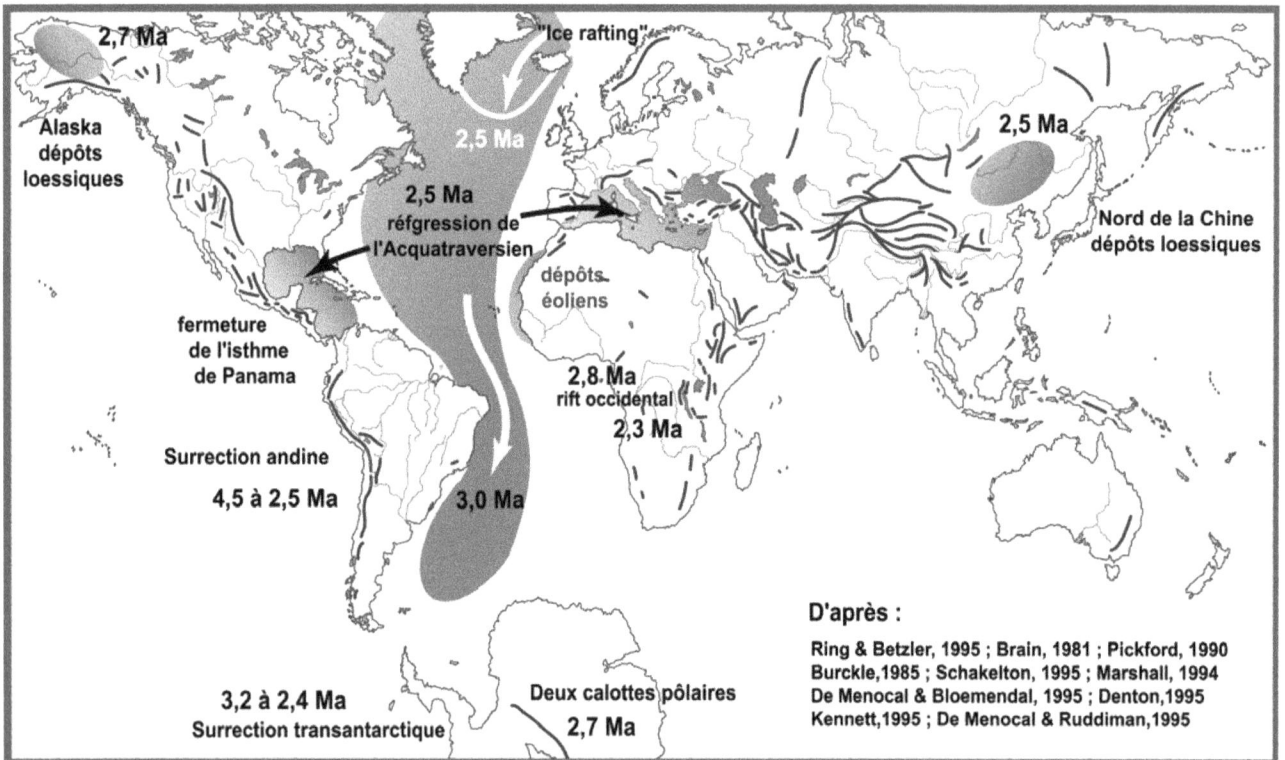

Figure 4 - Evénements climatiques et tectoniques globaux avant 2,5 Ma

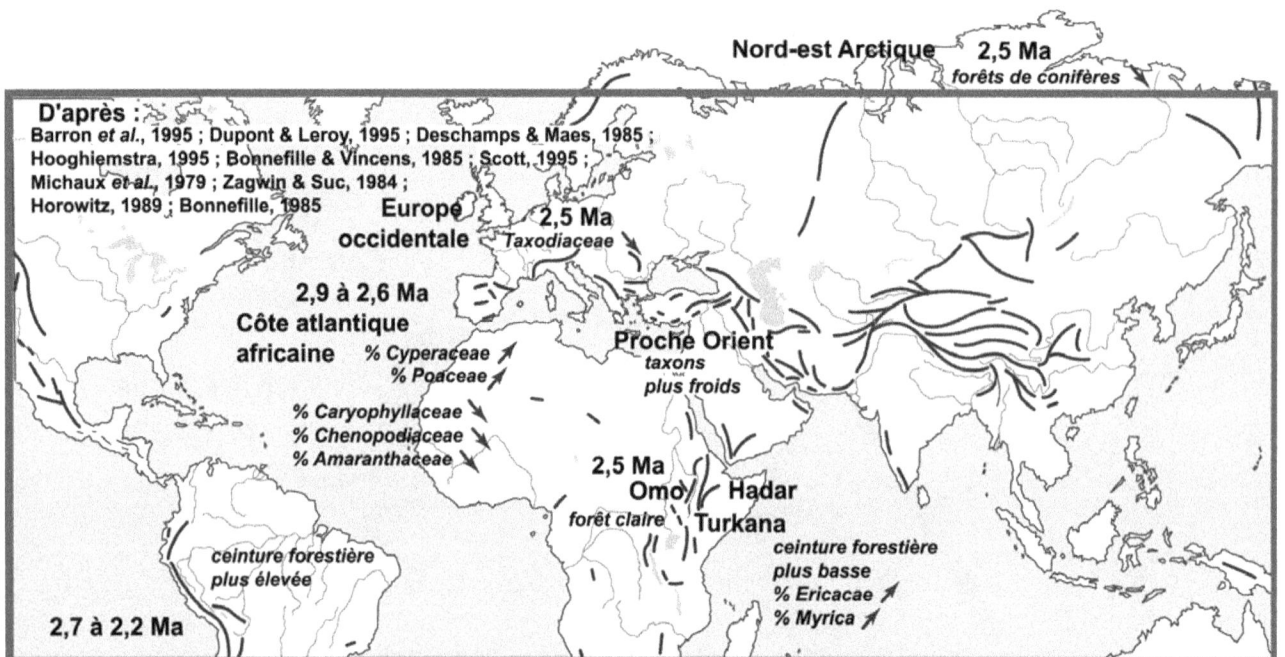

Figure 5 - Changements floristiques vers 2,5 Ma

principaux, documentés sur toute l'Afrique de l'est mais principalement à Hadar indiquent des conditions plus froides dans les hautes terres et des conditions plus sèches dans les basses terres avec, en Ethiopie, à Gaded, une descente altitudinale conséquente des ceintures forestières, le bassin Omo-Turkana étant lui, l'objet d'une période d'extension de la prairie. Au Proche Orient, les assemblages palynologiques sont marqueurs d'un refroidissement majeur. En Eurasie, en Sibérie et dans l'arctique nord-oriental, les forêts de conifères étaient présentes jusqu'à environ 2,5 Ma. Les *Taxodiaceae* voient leur extinction progressive ouest-européenne s'achever avec une toundra en Hollande et une association steppique de type quaternaire en Méditerranée. En Amérique du sud, à l'est de la Cordillère colombienne, la ligne forestière supérieure plus élevée qu'actuellement, est le signe de conditions climatiques plus froides.

La faune à 2,5 Ma

La faune est l'objet de nombreux bouleversements avec des taxons qui disparaissent quand d'autres apparaissent avec surtout de grandes migrations (figure 6).

Entre le lac Malawi et l'Afrique du sud une dispersion s'observe depuis le sud vers l'est, la diversité faunique maximale se situant après 2,5 Ma. L'étude des rongeurs indique que dans la formation Shungura, l'environnement est mésique à 70 % avec également des taxons indicateurs d'aridité. Les proboscidiens primitifs considérés comme forestiers disparaissent et sont remplacés par *E. recki brumpti* et *L. exoptata*. En Afrique de l'est et du sud, davantage d'espèces d'antilopes autres qu'Alcelaphinés et Antilopinés s'éteignent. En Afrique orientale, un changement apparent dans la composition des assemblages de Cercopithecidés se produit quand l'Afrique du nord, prise dans son ensemble, est marquée par la disparition de *Macaca* puis l'apparition de *Theropithecus*. Des changements fauniques se produisent également dans d'autres parties du monde ainsi une grande extinction des mollusques se produit en Méditerranée. Le

cheval *Equus* devenu prédominant dans les communautés fauniques en Amérique du nord voit des conditions de dispersions favorables peu de temps avant que *Synaptomys* - Lemming originaire d'Asie- n'apparaisse en Amérique. *Mammuthus* n'entre pas en Amérique, ni *Equus* en Afrique, mais leurs entrées sont synchrones en Europe où, les faunes rusciniennes sont effectivement modifiées, avec l'extinction de *Mammut borsoni, Tapirus arvernensis, Sus minor* et *Ursus minimus* et l'immigration de *Archidiskodon gromowi*, de *Gazella borbonica* et de *Equus cf livenzovensis*. La faune européenne s'appauvrit en musaraignes avec la première disparition de trois genres de soricidés en Europe centrale : *Sulimskia, Zelceina* et *Mafia* puis un quatrième genre *Blarinella* en Europe. En Inde, les Rhinocéros unicornes réapparaissent avec *Rhinoceros sivalensis*. Les Caprins et *Canis* apparaissent en Afrique depuis l'Eurasie. Le court intervalle centré autour de cette période témoigne de la disparition d'environ du tiers des vertébrés locaux, y compris les Mastodontes, Hipparions, Merycopotamides, plusieurs Suidés et la tortue géante et, l'apparition d'environ 20 nouveaux taxons qui constituent les deux tiers de l'assemblage faunique après 2,5 Ma, ainsi *Oryx, Damaop* et *Equus* remplacent *Hipparion*. Dans les Siwaliks, apparaissent des Cervidés alors qu'en Europe occidentale les Alcélaphinés et les Hippotraginés sont d'origine africaine. Par ailleurs, *Conepatus* et *Platyognus* migrent en Amérique du sud.

Les données géologiques à 1,8 Ma

La sécheresse générale décrite depuis 2,9 Ma jusqu'à un climax à 2,5 Ma, a des fluctuations avec une phase plus humide marquée vers 1,8 Ma (figure 7).

Les dépôts éoliens des carottes océaniques recueillis tant à l'est qu'à l'ouest de l'Afrique sont indicateurs d'un retour sec après cet épisode. Les dépôts éoliens recueillis au large de la côte nord atlantique de l'Afrique sont plutôt forts sauf vers 1,8 Ma quand les isotopes de l'oxygène suggèrent une réduction du couvert de glace. A cette date, des événements

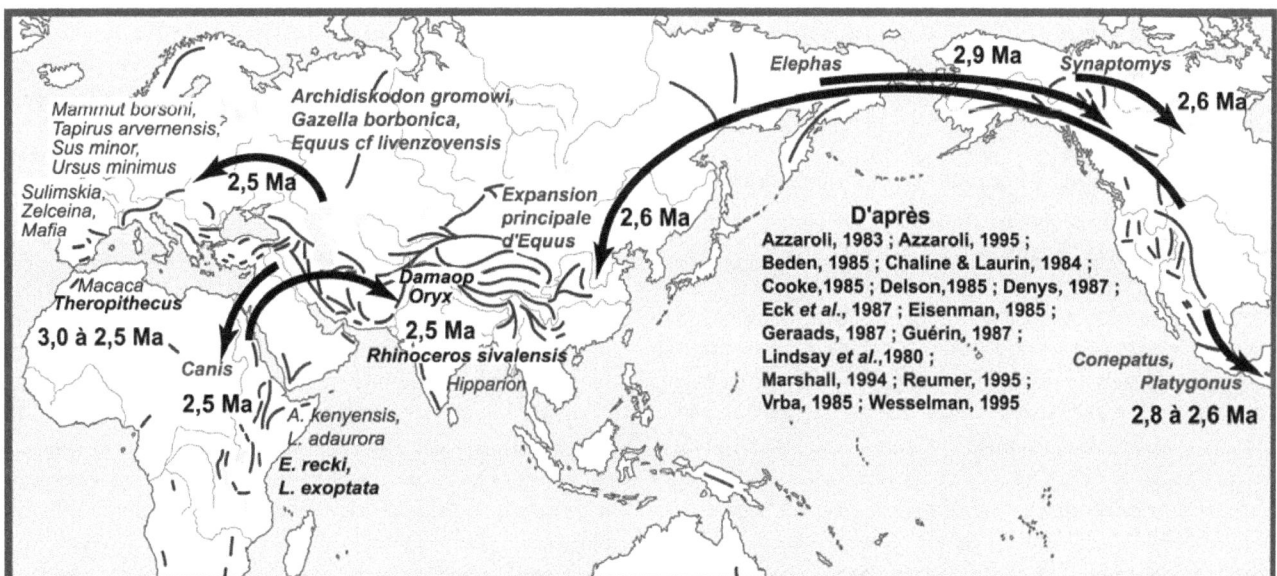

Figure 6 - Changements fauniques vers 2,5 Ma

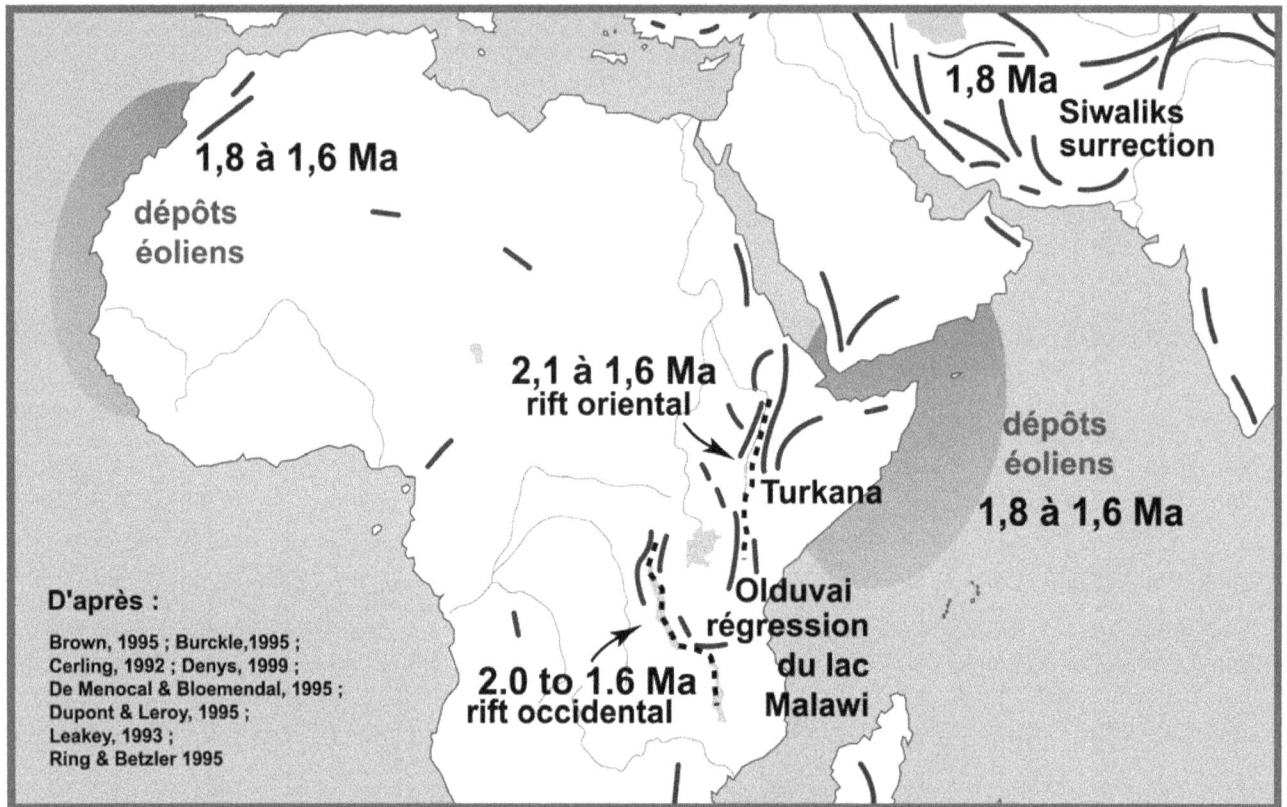

Figure 7 - Evénements géologiques avant 1,8 Ma

tectoniques se déroulent quand les dépôts d'Olduvai sont lacustres. La formation initiale d'un lac dans le bassin Turkana est également le fruit d'une construction volcano-tectonique. La phase fluviatile de dépôt se termine quand le cours de l'Omo est bloqué tectoniquement et qu'un grand lac stable inonde le bassin. Alors, prévaudrait une alternance de régimes fluviatiles et lacustres. Plus au sud, le lac Malawi subit une régression prononcée. Les données isotopiques des paléosols marquent des augmentations drastiques des d ^{18}O dans le bassin Turkana vers 1,8 Ma avant une augmentation du d^{13}C à 1,7 Ma. Les deux phénomènes sont synchrones à Olduvai. Enfin, en Eurasie, l'enregistrement géologique des Siwaliks prouve une forte surrection de la ceinture montagneuse.

La flore à 1,8 Ma

Le changement de la flore est-africaine est interprété soit comme des conditions plus arides après 1,8 Ma soit par une humidité à 1,8 Ma (figure 8).

Une phase aride existe à l'Omo et à l'Est Turkana : les groupements steppiques locaux et les groupements afromontagnards régionaux présentent des modifications importantes avec disparition complète des taxons arborescents soudano-zambéziens et une augmentation des pourcentages de Graminées liée à une extension des espaces herbacés en bordure de lac. Une extension de prairie est enregistrée à Koobi Fora. Au même moment sur les plateaux éthiopiens, se produit la substitution de groupements humides par des associations plus sèches. Plus au nord, à Hadar des

conditions de plus grande humidité ont lieu à 1,8 Ma. En Tanzanie, à 1,77 Ma la région d'Olduvai, marquée par l'aridité, voit sa végétation devenir progressivement plus boisée avec *Acacia*. Dans la vallée de l'Omo, la formation Shungura contient des bois silicifiés souvent roulés et remaniés, signes de conditions plus humides. En Amérique du sud, la végétation de paramo s'étend dans l'est de la Cordillère bolivienne avec installation d'un climat froid persistant.

La faune à 1,8 Ma

Il y a remplacement et disparition de nombreux taxons mais également des aires d'extension plus importantes pour certains autres en Europe, en Afrique et en Amérique (figure 9).

En Ethiopie, dans la formation Shungura, apparaissent des formes associées aux savanes très ouvertes : *Equus, Phacochoerus, Hippopotamus gorgops, Stylochoerus, Alcelaphinae*. Après cette évolution générale marquée par une ouverture des milieux se produit une mosaïque d'habitats où s'opposent artiodactyles non Bovidés / Bovidés et *Theropithecus* et Hominidés, les autres primates étant à part. *Ceratotherium simum* devient l'espèce dominante de Rhinocéritidé. Les Loxodontes disparaissent d'Afrique orientale, tandis que *Elephas recki* devient extrêmement polymorphe. Les faunes de l'Omo et d'Olduvai montrent la plus grande diversité avec dès 1,7 Ma un écosystème qui serait similaire à celui du Serengeti actuel. L'assemblage faunique micromammalien de la formation Shungura est

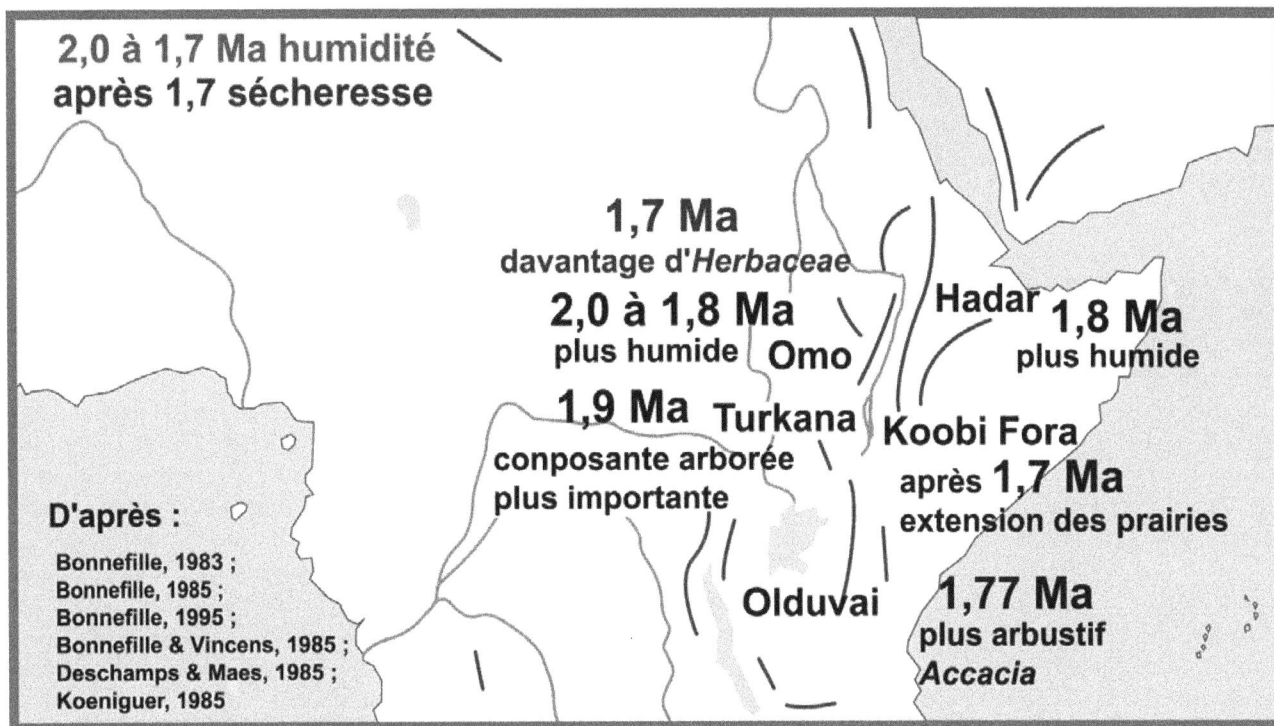

Figure 8 - Changements floristiques vers 1,8 Ma

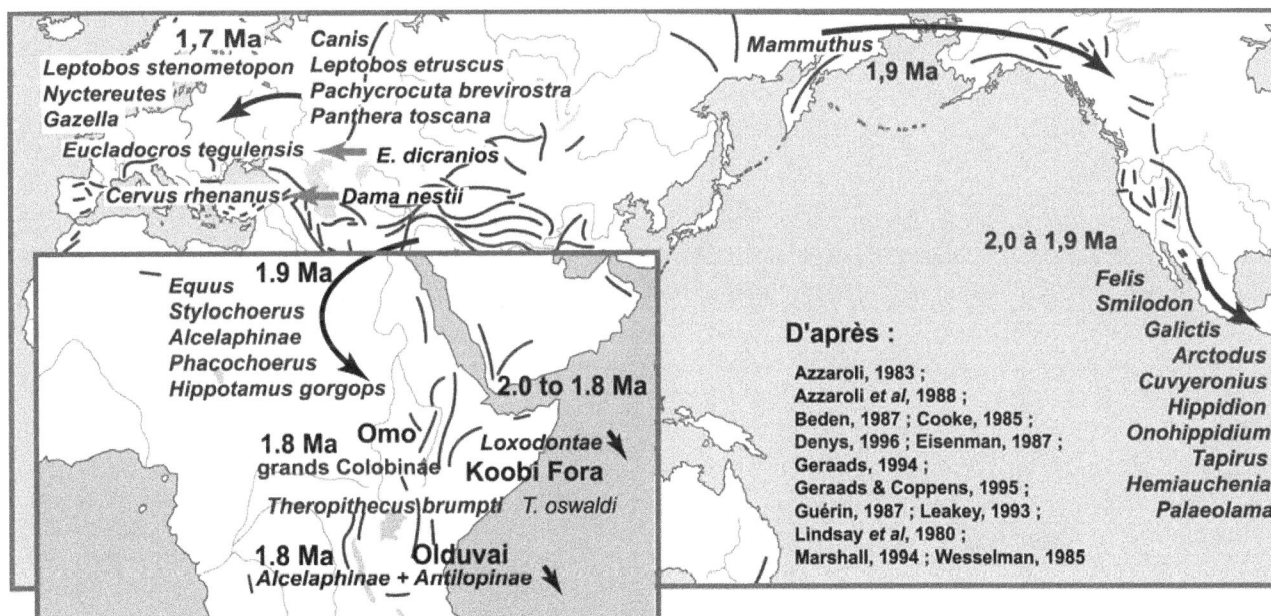

Figure 9 - Changements fauniques avant 1,8 Ma

xérique (86%). Concernant les bovidés, le taux d'Alcelaphiné + Antilopiné devient inférieur à 40% à Olduvai où s'exprime une baisse de la diversité des assemblages fauniques. Les primates sont également marqués par des changements, ainsi dans le bassin Turkana, *Theropithecus oswaldi* remplace *Theropithecus brumpti* comme le cercopithèque commun. En Eurasie, disparaissent *Nyctereutes, Gazella, Leptobos stenometopon*. Une expansion massive de *Canis etruscus* se produit en Europe avec également *Pachycrocuta brevirostra, Panthera toscana, Leptobos etruscus*. Alors, *Eucladocros*

tegulensis est remplacé par *E. dicranios* et, *Cervus rhenanus* par *Dama nestii*. Cet événement est marqué par le retour du loup asiatique en Europe. Toujours en Eurasie, les variations morphologiques entre populations de la lignée de *Mimomys occitanus-ostramosensis* traduisent une évolution ralentie. *Mammuthus* se répand en Amérique du nord et des événements migratoires importants ont lieu en Amérique du sud avec l'arrivée de *Felis, Smilodon, Galictis, Arctodus, Cuvyeronius, Hippidion, Onohippidium, Tapirus, Hemiauchenia* et *Palaeolama*.

DISCUSSION

Il résulte de ce bilan qu'il existe des indices indépendants et homogènes, quelque soit le champ disciplinaire dont ils proviennent, d'une sécheresse et d'un refroidissement s'exprimant à 2,5 Ma jusqu'à un regain d'humidité limité et documenté, tout au moins en Afrique orientale, à 1,8 Ma. Dans ce contexte, la radiation humaine qui se produit en Afrique orientale et qui est identifiée sur la seule base d'une étude anatomique, s'insert de manière cohérente dans l'intervalle de deux changements climatiques. Des radiations au sein des Bovidés et des Cercopithécidés - Primates vivant dans le même environnement que l'homme - sont clairement établies à cette époque (figure 10). Il semble donc légitime de considérer que des causes environnementales identiques aient été le moteur de mécanismes évolutifs similaires en particulier, au sein de différentes lignées de Primates.

Des travaux de synthèse (voir dans Zeitoun, 2000b) confortent d'ailleurs l'existence du synchronisme présent entre histoire évolutive des Hominidés et changements environnementaux. Une «haute» fréquence de spéciation au sein même du genre *Homo* traduite par une radiation n'est pas extravagante et ceci d'autant plus si l'on accepte que la quantité ou le taux de changement peut être particulièrement élevé pendant la phase initiale de spéciation dans des populations isolées de petite taille comme le suggèrent (Vrba, 1980 p. 66 et Foley, 1993). Concernant le genre *Homo*, il faut cependant indiquer qu'un changement culturel notable

a été décrit à l'origine de cette période, ou du moins à partir de 2,34 Ma dans l'ouest Turkana (Roche *et al.* 1999). Peut-être que des innovations techniques sont le marqueur du développement cognitif de certaines populations à moins, autre membre de l'alternative, que celles-ci aient contribué à une meilleure mobilité. Mais, le degré de définition des analyses typo-technologique lithiques ne permettent pas encore de préciser si les avancées technologiques sont le moteur ou la conséquence de l'évolution biologique à ce moment et dans cette aire géographique.

Pour ce qui est du retour partiel à l'humidité vers 1, 8 Ma, c'est peut-être là, par une modification des contraintes environnementales et des ressources naturelles dans plusieurs territoires, l'opportunité de la sortie de l'homme du continent africain. Les ancêtres d'*Homo erectus* ou d'*Homo sapiens* ou encore leur ancien cousin *Homo ergaster* comme le suggèrent les découvertes de Dmanisi en Géorgie pourraient être ces conquérants à l'origine des nouveaux modèles migratoires proposés (Turner, 1999 ; Bar-Yosef & Belfer-Cohen, 2001). Les migrations humaines les plus précoces hors d'Afrique auraient été déclenchées par les changements environnementaux de cette période selon une série d'événements isolés peut être également entremêlés d'extinctions de certaines populations.

Considérant l'hypothèse phylogénétique rapportée ici et les découvertes de terrain de ces dernières années, le constat de plusieurs vagues migratoires humaines spécifiques hors

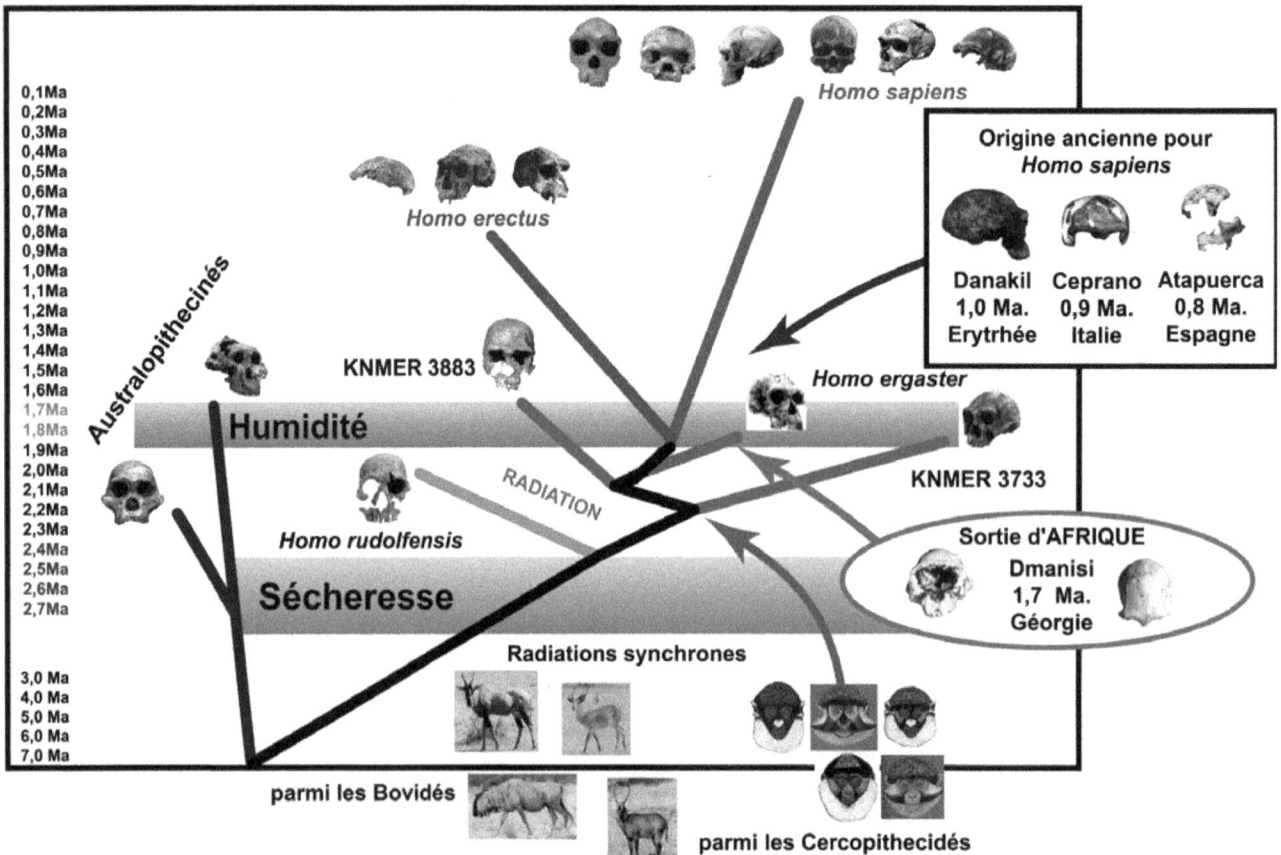

Figure 10 - Scénario évolutif conséquent de l'hypothèse phylogénétique présentée

d'Afrique apparaît. L'avènement des premiers *Homo* aurait eu lieu vers 2,4 Ma en Afrique orientale, suivi par une radiation précoce. Rapidement des *Homo erectus* sont reconnus en Afrique de l'est et en Indonésie, des *Homo ergaster* en Afrique de l'est et en Géorgie. La question de la pérennité de ces populations en Eurasie n'est pas traitée ici. Toujours est-il, qu'à partir d'un million d'années apparaîtraient les premiers représentants d'*Homo sapiens*, toujours en Afrique de l'est, en Erythrée. Se répandant vers le sud et le nord de l'Afrique puis au-delà, ces populations acquièrent alors partout et rapidement des spécificités régionales. Les premiers *Homo sapiens sapiens* seraient identifiés vers 300 000 ans en Afrique avec le spécimen éthiopien de Bodo et, déjà au Proche-Orient vers 250 000 avec Zuttiyeh (Zeitoun, 2001). Ce n'est que vers 35 000 ans qu'ils atteignent l'Europe et l'Australie. Et, depuis lors, fait original, ils sont les seuls survivants du genre humain. Une alternative à ce remplacement complet et global par une sous-espèce hégémonique est d'envisager que, sous sa forme actuelle, l'homme moderne est l'expression de l'homogénéisation du pool génique d'un réservoir global unique, quand, au Pléistocène, les différents domaines régionaux étaient suffisamment isolés et/ou le nombre d'individus trop faible pour ce pool génique soit, à l'époque, compartimenté en sous-espèces distinctes. Par cette approche et selon ce schéma récapitulatif, les deux modèles « Sortie d'Afrique » et « Evolution multirégionale » sont réconciliés ou plus exactement, n'ont plus lieu d'exister, puisque tout se passe au niveau sous-spécifique ! Cependant, à partir d'un million d'années, la question reste posée de la pérennité plus ou moins prolongée et étendue géographiquement des premiers humains hors d'Afrique. En effet, si en Afrique une population d'*Homo ergaster* ou *Homo erectus* a été l'ascendant d'*Homo sapiens,* qu'en est-il des autres populations ? L'hypothèse de leur extinction apparaît comme la plus probable au regard de la biologie, la spéciation étant considérée comme un phénomène historiquement mais également géographiquement unique.

CONCLUSION

Un moyen de tester la congruence entre événements environnementaux et changements évolutifs consiste à mettre en regard les structures de parenté des taxons et les événements environnementaux dans un cadre chronologique. Ainsi pourront-ils être traduits comme une explication des processus évolutifs. A partir d'une analyse phylogénétique basée sur l'étude anatomique des spécimens sans regroupement arbitraire, une redéfinition d'*Homo erectus* conduit à poser l'hypothèse d'une radiation africaine, d'une expansion humaine ainsi que de l'avènement d'*Homo sapiens*, les trois phénomènes étant précoces. Une congruence entre les données paléontologiques et environnementales inférées par l'enregistrement fossile apparaît. La reconnaissance analytique des anciens humains comme des *Homo sapiens* et non des *Homo erectus* permet de concilier les deux modèles « sortie d'Afrique » et « Evolution multirégionale » ou du moins de démontrer que c'est un faux problème à ce niveau. Ainsi, *Homo sapiens* sort d'Afrique et, comme le suggère la biologie, une continuité morphologique

régionale est plus aisément acceptable entre des sous-espèces. Si ce travail peut être un argumentaire pour proposer un nouveau paradigme, il s'agit de le mettre à l'épreuve des données nouvelles comme c'est le jeu en science.

Adresse de l'auteur

V. ZEITOUN
UPR 2147 du CNRS « Dynamique de l'évolution humaine »
44 rue de l'amiral Mouchez, 75 014 Paris FRANCE

Bibliographie

ABBATE, E., ALBIANELL,I A., AZZAROLI, A., BENVENUTI, M., TESFAMARIAM, B., BRUNI, P., CIPRIANI, N., CLARKE, R.J., FICCARELLI, G., MACCHIARELLI, R., NAPOLEONE, G., PAPINI, M., ROOK, L., SAGRI, M., TECLE, T.M., TORRE, D., VILLA, I., 1998, A one million year old Homo cranium from the Danakil (Afar) Depression of Eritrea. *Nature* 393, p. 458-460.

ASCENZI, A., MALLEGNI, F., MANZI, G., SEGRE, A. & SEGRE NALDINI, E., 2000, A re-appraisal of Ceprano calvaria affinities with *Homo erectus*, after the new reconstruction. *Journal of Human Evolution* 39, p.443-450.

BAR-YOSEF, O. & BELFER-COHEN, A., 2001, From Africa to Eurasia – early dispersals. *Quaternary International* 75, p. 19-28.

BERMUDEZ DE CASTRO, J.M., ARSUAGA, J.L., CARBONELL, E., ROSAS, A., MARTINEZ, I., MOSQUERA, M., 1997, A Hominid from the Lower Pleistocene of Atapuerca, Spain : Possible ancestor to Neandertals and Modern Humans. *Science* 276, p.1392-1395.

BRAUER, G. & MBUA, E. 1992, *Homo erectus* features used in cladistics and their variability in Asian and African hominids, *Journal of Human Evolution* 22, p. 79-108.

CAMPBELL, B., 1972, Conceptual progress in physical anthropology : fossil man. *Annual Review of Anthropoly* 1 p.27-54.

DUBOIS, E., 1940, The fossil humain remains discovered in Java by Dr. G.H.R. von Koenigswald and attributed by him to *Pithecanthropus erectus*, in reality remains of *Homo sapiens soloensis*. (continuation). *Nederl. Akademie van Wetenschappen* 43 p. 841-854.

FALGUÈRES, C., YOKOYAMA, Y., JACOB, T., & SÉMAH, S., 2001, Advancements in the dating of Solo Man. Actes du XIVème Congrès de l'Union Internationale des sciences préhistoriques et protohistoriques, Lièges.

FOLEY, R., 1993, African terrestrial primates : the comparative evolutionary biology of *Teropithecus* and the Hominidae. In *Theropithecus the rise and fall of a primate genus* edited by N. Jablonski. Cambridge University Press, p. 245-270.

GABUNIA, L., & VEKUA, A., 1995, A Plio-Pleistocene hominid from Dmanisi, East Georgia, Caucasus. *Nature* 373, p. 509-512.

GROVES, C., 1989, *A theory of human and primate evolution.* Oxford : Oxford University Press.

HOWELL, F., 1999, Paleo-demes, species clades, and extinctions in the pleistocene hominin record. *Journal of Anthropological Research* 55, p. 191-253.

KENNEDY, K., SONAKIA, A., CHIMENT, J. & VERMA, K., 1991, Is the Narmada hominid an Indian *Homo erectus* ?, *American Journal of Physical Anthropology* 86, p. 475-496.

LIEBERMAN, D., WOOD, B., & PILBEAM D., 1996, Homoplasy and early *Homo* : an analysis of the evolutionary relationships of *Homo habilis sensu stricto* and *Homo rudolfensis. Journal of Human Evolution* 30, p. 97-120.

ROCHE, H., DELAGNES, A., BRUGAL, J.P., FEIBEL, C., KIBUNJIA, M., MOURRÉ, V. & TEXIER, P.J., 1999, Early hominid stone tool production and technical skill 2.34 Myr ago in West Turkana, Kenya. *Nature* 399, p. 57-60.

STRINGER, C., 1986, The credibility of *Homo habilis.* In *Major Topics in Primate and Human Evolution edited by* : B. Wood : Cambridge University Press, p. 266-294.

STRINGER, C., 1987. A numerical cladistic analysis for the genus *Homo. Journal of Human Evolution* 16, p. 135-146.

SWISHER, C., RINK, W., ANTON, S., SCHWARCZ, H., CURTIS, G., SUPRIJO, A., WIDIASMORO, 1996, Latest *Homo erectus* of Java : potential contemporaneity with *Homo sapiens* in southeast Asia. *Science* 274, p. 1870-1874.

TOBIAS, P., 1985, Single characters and total morphological pattern redefined the sorting effected by a selection of morphological features of the early hominids. In *Ancestors : the hard evidence* edited by A. Liss, p. 94-101.

TURNER, A., 1999, Assessing earliest human settlement of Eurasia : Late Pliocene dispersions from Africa. *Antiquity* 73, p. 563-570.

VRBA, E., 1980, Evolution, species and fossils : how does life evolve ? *South African Journal of Science* 76, p. 61-84.

WANPO, H., CIOCHON, R., YUMIN, G., LARICK, R., QIREN, F., SCHWARCZ, H., YONGE, C., DE VOS, J. & RINK, W., 1995, Early *Homo* and associated artefacts from Asia. *Nature* 378, p. 275-278.

ZEITOUN, V., 2000a, Révision de l'espèce *Homo erectus* (Dubois, 1893), utilisation des données morphologiques et métriques en cladistique, *Bull. et Mém. de la Société d'Anthropologie de Paris, mémoire spéciale* 1, p. 1-200.

ZEITOUN, V., 2000b, Adéquation entre changements environnementaux et spéciations humaines au Plio-Pléistocène. *C. R. Acad. Sci.* 330, p. 161-166.

ZEITOUN, V., 2001, The taxinomical position of the skull of Zuttiyeh. *C. R. Acad. Sci.* 332, p. 521-525.

ZHOU, M., LI, Y, & WANG, L., 1982, Chronology of the chinese hominids In *L'Homo erectus et la place de l'Homme de Tautavel parmi les hominidés fossiles*, 1er Congrès international de paléontologie humaine, prétirage, edited by H. Lumley H de, p. 593-604.

BOCKSTEIN (GERMANY) – A NEW LOOK AT AN OLD PROBLEM
A PRELIMINARY REPORT

Petra KRÖNNECK

Résumé : La grotte de Bockstein a été fouillee pendant les années 30 et 50 par Robert Wetzel. Les restes ostéologiques des campagnes des années 30 ont été analysés par R. Wetzel et U. Lehmann et considérés en première ligne sous leur aspect paléontologique. Cet article veut présenter quelles importantes possibilités d'analyse nous offre le matériel de ces deux campagnes si on le considère du point de vue archéozoologique.

Abstract: The Bockstein cave system in southern Germany was excavated in the 1930s and the 1950s by Robert Wetzel. The faunal remains of the 1930s campaign were analysed by R. Wetzel and U. Lehmann. They, however, considered the bones under a palaeontological aspect. In this paper, it is shown, that we can get much more possibilities of analyses when we consider the assemblages of both campaigns under an archaeozoological aspect.

The valley of the River Lone is located in southern Germany northwest of the city of Ulm; it extends approximately 40 kms in a northwestern direction. The valley is well known for its Paleolithic findspots such as the Bärenhöhle, which was already explored in 1862 by Oskar Fraas, or the Vogelherd Cave with its famous *art mobilier*.

The Bockstein cave system is situated at the lower course of the river and forms a cluster of small caves and abris which extent on the northern edge of the prealpine molasse basin, only a few kilometers north of the River Danube. All elements of this cave system are located on the western slope of the valley so their present-day entrances are nowadays exposed to storm and rain.

As early as the second half of the 19th century the first excavations at Bockstein were conducted by Ludwig Bürger. In 1879 and again in 1883/84 he explored one part of the cave and abri system (the Bocksteinhöhle). It was here where he not only discovered diverse layers of the Pleistocene but he also found one of the rare mesolithic burials which are presently known in Central Europe. Unfortunately, most of the finds recovered by Bürger are lost today (WEHRBERGER 2000).

In 1908 R.R. Schmidt, who later founded the *Institut für Urgeschichte* at the University of Tübingen, started some small-scale excavations. The results of his work which were published in 1912 could corroborate the stratigraphy Bürger had already recognized.

It was the anatomist and archaeologist Robert Wetzel who started profound excavations in 1932. His first camp aign which lasted until 1935 was followed by a second almost 20 years later (1953 – 1956). During this time he investigated two small caves (Bockstein-Loch and Bockstein-Westloch) with their entrance areas and several abris. He also re-opened the original Pleistocene entrance

of that cave, which Bürger had already excavated more than 50 years before.

Parallel to the excavations in the 1950s Wetzel and others began to analyse the botanical remains, the sediments as well as the animal bones. Regrettably it is true that only the bones of the 1930s campaign were analysed whereas the bones found in the 1950s remained unprocessed. The results of these efforts were finally published in 1969 (WETZEL/BOSINSKI 1969).

Due to this comparatively well published situation the Bockstein Cave was chosen to begin new examinations of the faunal remains. Two aspects should be emphasized. On the one hand it is possible to compare two different methods of evaluation, the palaeontological and the archaeozoological one, by using the bone material of the 1930s campaigns. On the other hand the extensive analysis of the faunal remains of both campaigns renders it possible to get a good survey of the climatic development of larger parts of the Upper Pleistocene.

It is also possible to compare the results of archaeozoological examinations with those of botanical and sedimentological analyses made by Paul Filzer and Elisabeth Schmidt in the 1950s (WETZEL/BOSINSKI 1969). Indeed, R. Wetzel has also started some taphonomic researches, but unfortunately, he was unable to finish this before his death in 1962.

Although Wetzel and Lehmann performed a vast number of analyses they restricted themselves to solely qualitative aspects when coming to their conclusions. The relationships between the species were determined by using the number of identified specimens (NISP) and the minimum number of individuals (MNI). It was H.-P. Uerpmann, who pointed out, however, that it is more helpful to compare the weight of the bones of each species (UERPMANN 1972). Following this proposal, it is possible to evaluate the nutritional importance of each species in more detail.

Palaeontology by Ulrich Lehmann
number of identified specimens
minimum number of individuals (MNI)

skeletal elements
(age & sex)

Archaeology by Robert Wetzel
gnawing marks
cut marks

Archaeozoology by Petra Krönneck
number of identified specimens

weight
skeletal elements
age & sex

gnawing marks
cut marks

Different methods of palaeontology and archaeozoology used for faunal remains in Bockstein III.

In order to demonstrate the differences between the two methods we use a Middle Palaeolithic layer of Bockstein-schmiede/loch (Bockstein III). The stone artifacts found in this layer enable us to classify it as belonging into the Keilmesser complex.

The best example for our comparison is unit 21 in Bockstein III. At first sight the number of unidentified fragments seems to be overwhelming; however, if we take, the weight into account it is obvious that this category does not have such an important significance. It rather shows that the category of the unidentified bones consists of tiny fragments of an average of 3.8 g each.

Looking at the importance of the megafauna which includes *Coelodonta antiquitatis* and *Mammuthus primigenius* we get the opposite situation. Although we just have a few fragments of these species the weight tells us that they seem to represent a considerable contribution to the daily menu (fig. 1).

But of course, weight measurements on their own can not be used for reconstructing the faunal sustenance; we have to consider the skeletal elements as well. Referring for example to the mammoth, it is a complete ulna and some molar fragments that add to the weight.

As we can see in picture 2 the *equidae* as well as the *bos/bison* – group are largely represented by teeth or teeth fragments. It goes without saying that these parts of an animal can not be regarded as crucial for appeasing hunger. But as these species are represented by all parts of their bodies, we do not have to think of a deliberate selection of teeth. Of course, on the other hand it can not be denied that horse ribs are evidently over-represented. Since they, however, represent acceptable amounts of meat as well as raw material for artifacts these figures can be used in both analyses.

Another remarkable example is represented by the remains of *Rangifer tarandus*. Although we can also confirm the presence of all parts of the skeleton the diagram clearly shows an abnormal proportion of metatarsal bones. It is certain that this situation can not be explained merely by reindeer´s being of great nutritional importance because not a lot of meat is

obtainable from this body part. It is, however, known that metatarsal bones were often used to produce artifacts. Indeed, all bone artifacts found in Bockstein IIII are made from reindeer metatarsal bones. Consequently, we have to consider this special situation when we try to assess the right position of *Rangifer tarandus* on the Palaeolithic bill of fare.

In order to produce a ranking list for the species we have to take into account all the aspects mentioned above. If we do so, we can give the following result:

1. Equidae

2. *Bos/Bison* and *Bison priscus*

3. Rangifer tarandus

The importance of *Mammuthus primigenius* is not easy to determine. The ulna mentioned above shows some wear of using so that it is probable that it just served as an artifact and had no relevance for sustenance.

The analysis of gnaw and cut marks which ... just carried out under describing aspects shows some interesting results. Although the evaluation is not finished yet it can be stated that in some cases the gnaw marks superimpose the cut marks. So we can be sure that carnivores chewed on those rubbish bones discarded off by men. This observation makes it extremely difficult to distinguish between anthropogenic and carnivorous prey. However, it is this question which is of great importance for a vast number of Palaeolithic sites.

Although not emphasized here the main aim of the thesis is, however, the evaluation of the climatic development from the Middle Palaeolithic in the Early Upper Pleistocene and the Aurignacian layers to the end of the Pleistocene with its Magdalenian remains.

In order to use faunal remains for reconstructing climatic and ecological conditions it is certainly necessary to use the correct ranking of the identified species. Therefore, it is necessary to make allowance for all the restrictive elements, some of which were mentioned above. Only then it is possible to construct serious statements about the meaning of the faunal assemblage for the Palaeolithic climate and ecology.

It is not possible to go into details here, but we are able to present the results for the Bockstein III – layer (Keilmesser –

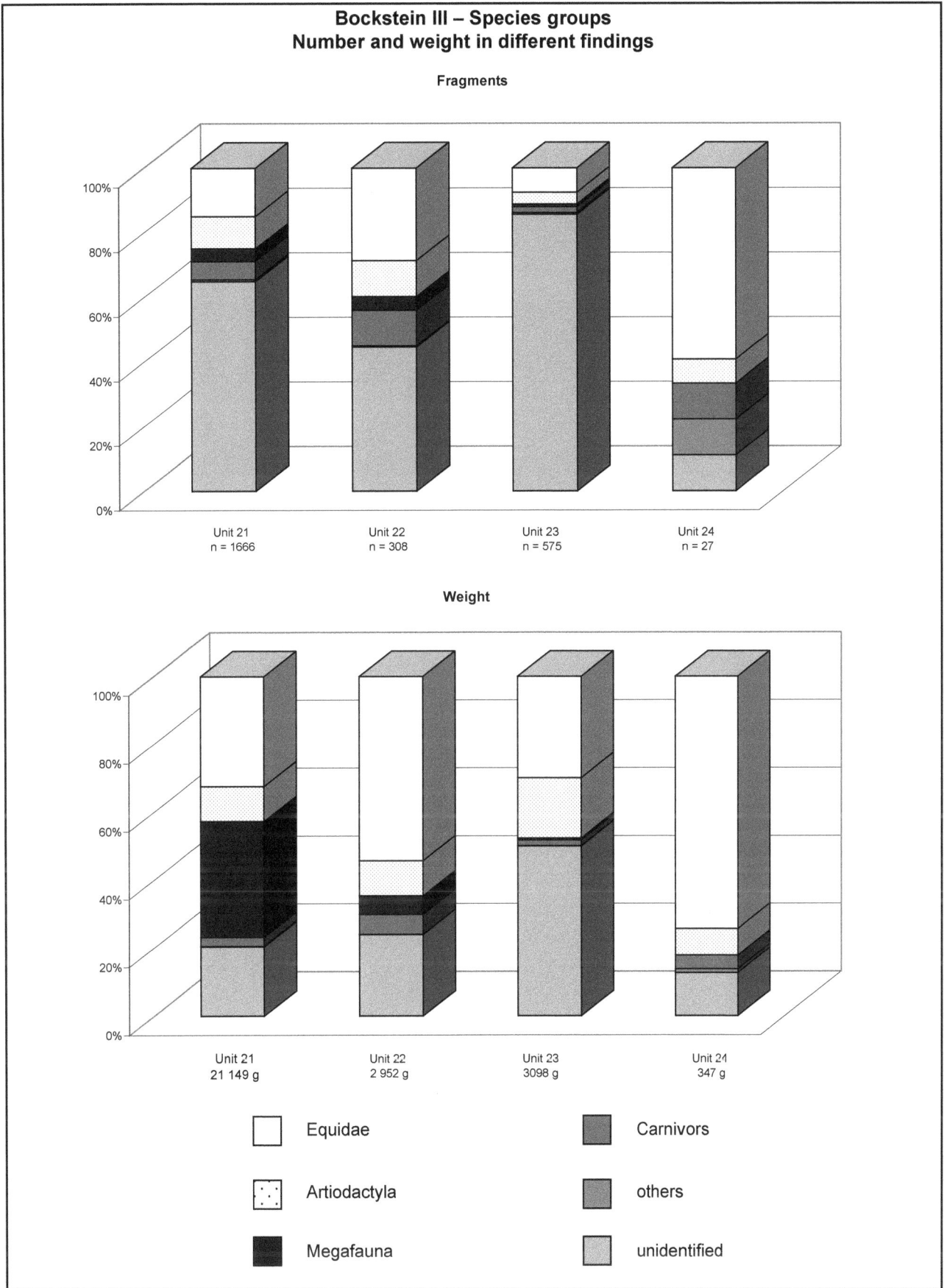

Bockstein III – Species groups
Number and weight in different findings

Figure 1.

complex) in rough outlines. Therefore, it is possible to reconstruct a forest steppe with moister habitats in the lowlands of the molasse basin as well as in the Lone valley itself. In contrast, the highlands of the Swabian Alb which were built to a great extent by a karstic jurassic landscape represent drier steppes.

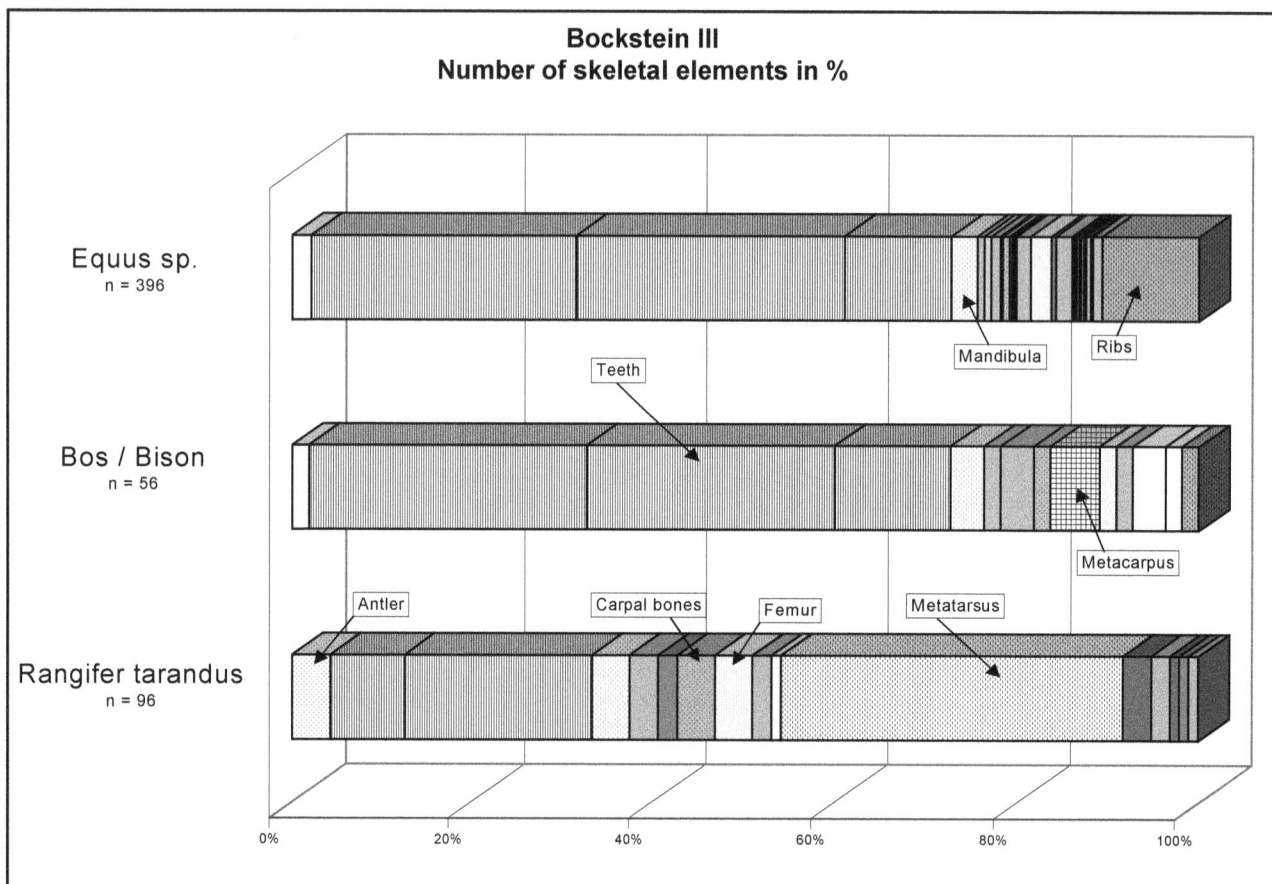

Figure 2.

This reconstruction based on the list of species (which is much more extensive than those of Lehmann's) fits very well into the picture of botanical and sedimentological analyses. Hence, we can assume a mosaic of woods and grassland on the northern spurs of the molasse basin between the Danube valley and the Swabian Alb.

In this short paper it was demonstrated that the Bockstein cave system, although excavated and already published a considerable time ago, holds quite a significant potential for further scientific research. Especially the faunal remains allow us plenty of possibilities of analyses when considered under an archaeozoological and not instead of a palaeontological aspect. Faunal remains can help us not only to get a better idea of what the faunal sustenance was like but, they also help us to find answers to questions concerning the climatic or the ecological situation.

Acknowledgements

The author thanks Kurt Wehrberger who kindly gave the permission to study the material. Special thanks go to Prof. Dr. Dr. Hans-Peter Uerpmann and Prof. Nicholas J. Conard Ph.D. for their constant support. I also want to thank K.-D. Dollhopf who provided valuable comments on an earlier draft of this paper. Last but not least, many thanks to Tina Jakob who kindly corrected the English version of this paper.

Notes

This paper is a short preliminary report of my thesis which will be written at the Institut of Pre- and Protohistory and Mediaeval Archaeology at the University of Tuebingen (Prof. Dr. Dr. Hans-Peter Uerpmann, Prof. Dr. Nicholas J. Conard)

Author's address

Petra KRÖNNECK
Inst fur Ur- und Frühgeschichte
Eberhard-Karls-Universität Tübingen
Burgsteige, 11, D-72070, Tübingen, ALLEMAGNE

Bibliography

UERPMANN, H.-P., 1972. Tierknochenfunde und Wirtschaftsarchäologie. Eine kritische Studie der Methoden der Osteo-Archäologie. *Archäologische Informationen* 1, 1972, 9-27.

WETZEL, R & BOSINSKI, G., 1969. Die *Bocksteinschmiede im Lonetal (Markung Rammingen, Kr. Ulm)*. Veröffentlichungen des Staatlichen Amtes für Denkmalpflege A 15 Stuttgart.

WEHRBERGER K., 2000. "Der Streit ward definitiv beendet..." Eine mesolithische Bestattung aus der Bocksteinhöhle im Lonetal, Alb-Danau-Kreis. *Archäologisches Korrespondenzblatt* 30, 2000, 15-31.

LATE MESOLITHIC LIVING SPACES
IN THE FRAMEWORK OF THE BLACK AND AZOV SEA STEPPE

Olena V. SMYNTYNA

Résumé : Cinq domaines identifiés comme "espace résidentiel" on peut distinguer dans les limites de la Boréal Steppe Ukraine. Ces espaces sont caractérisés par les traits spécifiques de la culture ethnique de leurs habitants de même que par les particularités de leur subsistance et mode de la vie. Le problème de l'implication ethnique de la pluralité de modes de l'exploitation des espaces résidentielles est avancé.

Abstract: Five regions identified as "living spaces" are distinguished in the frameworks of the Boreal Steppe zone of contemporary territory of Ukraine. These living spaces are characterised by specific features of material and ethnic culture of their inhabitants as well as by peculiarities of their economy and mode of life. The problem of ethnic implication of these living spaces exploitation plurality is put forward.

INTRODUCTION

Traditionally the transition to Late Mesolithic is correlated with Boreal period of Holocene. Substantial changes of palaeogeographic situation have happened at that time in the frameworks of Steppe Ukraine. The process of softening and, first of all, humidifying of the climate of this region has been reflected in fauna and vegetation changes. Steppe floral complex has become more mesophilous and could be interpreted as meadow steppe one. Among the steppe faunal inhabitants the proportion of species well adapted to semiclosed and closed biotopes has been significantly augmented. In these new palaeogeographic conditions steppe population mode of life also changes. Gradual increase of population density has become the general trend of steppe demographic situation development in Late Mesolithic. At that time quantity of steppe short-term sites and seasonal settlements has been uprising, and new types of archaeological sites (base camps and cemeteries) have appeared at that region [Smyntyna. 1999].

The change of settlement system basic principles implicates definite stabilisation of Steppe Late Mesolithic population mode of life, which, in its turn, was caused by general increase of biomass density per unit area. At the same time short-term sites introduced by small amount of picked-up flint artefacts still dominating in the Steppe inhabitants' settlement system. So, the Late Mesolithic demographic situation is appeared to be rather patchy one, demanding distinct distribution of food sources. As a result, the necessity of foraging territory clear demarcation has become the issue of principal importance. Archaeologically this phenomenon could be traced in on so-called "clusters" of short-term sites, situated around the central place (base camp) and attributed to the same culture.

Therefore, Late Mesolithic Steppe natural geographic zone, taken as a whole, is characterised by definite common features of culture and household activities of its inhabitants.

Nevertheless more detail spatial analysis of separate environmental components and population livelihood systems gives chance to reveal some important local peculiarities in steppe inhabitants' subsistence ways as well as in their ethnic culture. As a result at least five living spaces could be distinguished in the framework of Steppe Ukraine in Late Mesolithic times: Dnister-Dnipro interfluve, Lower Danube basin, Dnipro Rapids region, Crimea and Azov Steppe, Sivers'ky Donets' basin (Fig. 1).

DNISTER-DNIPRO INTERFLUVE

As well as in Early Mesolithic, during the Boreal period of Holocene this region embodies all the most typical traits of Steppe palaeogeography situation and demonstrates on this basis most typical features of steppe living space exploitation. Moreover, it was just given space where the continuity of livelihood system and ethnic culture evolution has been displayed in the fullest measure.

There are certain limitations of source base for reconstruction of this region fauna and flora as far as palinological analysis was not provided at the Late Mesolithic sites. Available now spore-pollen diagrams of Early Holocene subaeral alluvial sediments as well as those of swamps and estuaries have no absolute chronology. So the clear shift between Preboreal and Boreal periods of Holocene could not be proved on their base (Fig. 2). Nevertheless they indicate that in Late Mesolithic open steppe landscape was dominating on the broad space of Dnister-Dnipro interfluve. Proportion of grasses in vegetation structure sometimes reaches 80-90 %, among them species adapted to arid climatic conditions absolutely prevailed: proportion of *Chenopodiceae*, *Artemisiae* and *Graminae* taken together could be as much as 40-70 % of all grasses. Most widespread (if not only) faunal inhabitants here are horse *(Equus caballus L.)* and ass *(Asinus hidruntinus L.)*, bones of which are found at Late Mesolithic settlements of this locale (Table 1).

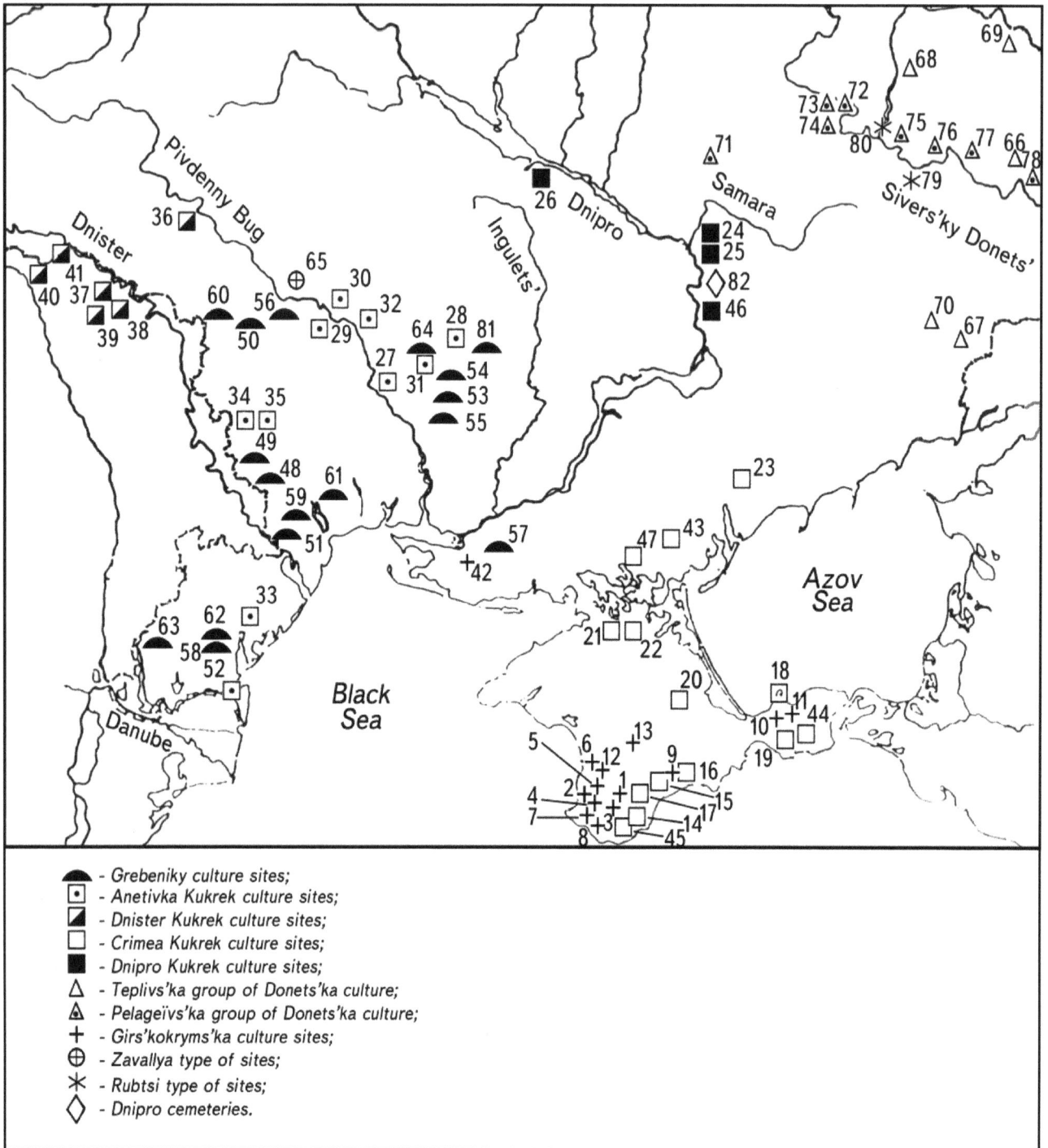

Legend:
- Grebeniky culture sites;
- Anetivka Kukrek culture sites;
- Dnister Kukrek culture sites;
- Crimea Kukrek culture sites;
- Dnipro Kukrek culture sites;
- Teplivs'ka group of Donets'ka culture;
- Pelageïvs'ka group of Donets'ka culture;
- Girs'kokryms'ka culture sites;
- Zavallya type of sites;
- Rubtsi type of sites;
- Dnipro cemeteries.

Figure 1. Late Mesolithic Living Spaces of Steppe Ukraine

1 – Fat'ma Koba, 2- Murzak-Koba, 3 – Snak-Koba, 4 – Kara-Roba, 5 – Zamil'-Koba, 6 – Alimovskiy naves, 7 – Laspi VII, 8 – Era 1,
9 – Adzhi-Koba III, 10 – Frontove I, 11 – Lenins'ke I, 12 – Tash-Ayir I, 13 – Kukrek, 14 – Balin-Kosh, 15 – Alabach, 16 – Su-At,
17 – Alatchuk, 18 – Tasunove, 19 – Lugove, 20 – Oleksiyivs'ka Zasukha, 21 – Dolynka, 22 – Ishun', 23 – Kam'yana Mohula, 24 – Igren'8,
25 – Kizlevyi, 26 – Belyka Andrusivka, 27 – Abuzova Balka, 28 – Sagaidak, 29 – Konetspol', 30 – Synuykhin Brid, 31 – Bubinka,
32 – Sofiyivka, 33 – Trapivka, 34 – Priymove, 35 – Karabanove, 36 – Gvozdeve, 37 – Varvarivka, 38 – Gura Kam'yanka, 39 – Frumushika,
40 – Stari Bedradzhi, 41 – Burlakiv Yar, 42 – Velyki Koponi, 43 – Vesnyanka VI, 44 – Gornostaivka, 45 – Kizil-Koba, 46 – Terlyans'kiy Yar,
47 – Sergiivka, 48 – Grebeniki, 49 – Girzheve, 50 – Olenivka, 51 – Dobrozhany, 52 – Myrne, 53 – Pelageivka, 54 – Balakha, 55 – Lenino II
and III, 56 – Poznanka, 57 – Velyki Kopani 1a, 58 – Borysivka, 59 – Karpove, 60 – Slobodka, 61 – Skosarivka, 62 – Tatarbunary,
63 – Zaliznichne, 64 – Sofiivka, 65 – Zavallya, 66 – Teple, 67 – Mospine, 68 – Prishib, 69 – Pelageivka III, 70 – Krem'yana Gora,
71 – Shevchenko, 72 – Petrovs'ke IV, 73 – Petrovs'ke X, 74 – Petrovs'ke XXVIII, 75 – Drobysheve, 76 – Petrovoorlovs'ka,
77 – Borovs'ke, 78 – Kondryuts'ka, 79 – Raigorodok, 80 – Rubtsi, 81 – Kazanka, 82 – III Vasylivs'kiy cemetery.

At the same time, in all without exception spectra definite increase of arboreal vegetation proportion is increasing in comparison with previous period. In this category of flora the percentage of deciduous species is considerable augmenting. So, proportion of oak *(Quercus sp.)*, elm *(Ulmus sp.)*, linden *(Tilia)* and other deciduous species, taken together, sometimes riches 25 % of total amount of tree pollen. Nevertheless pine (Pinus silvestris) and birch (Betula sp.)

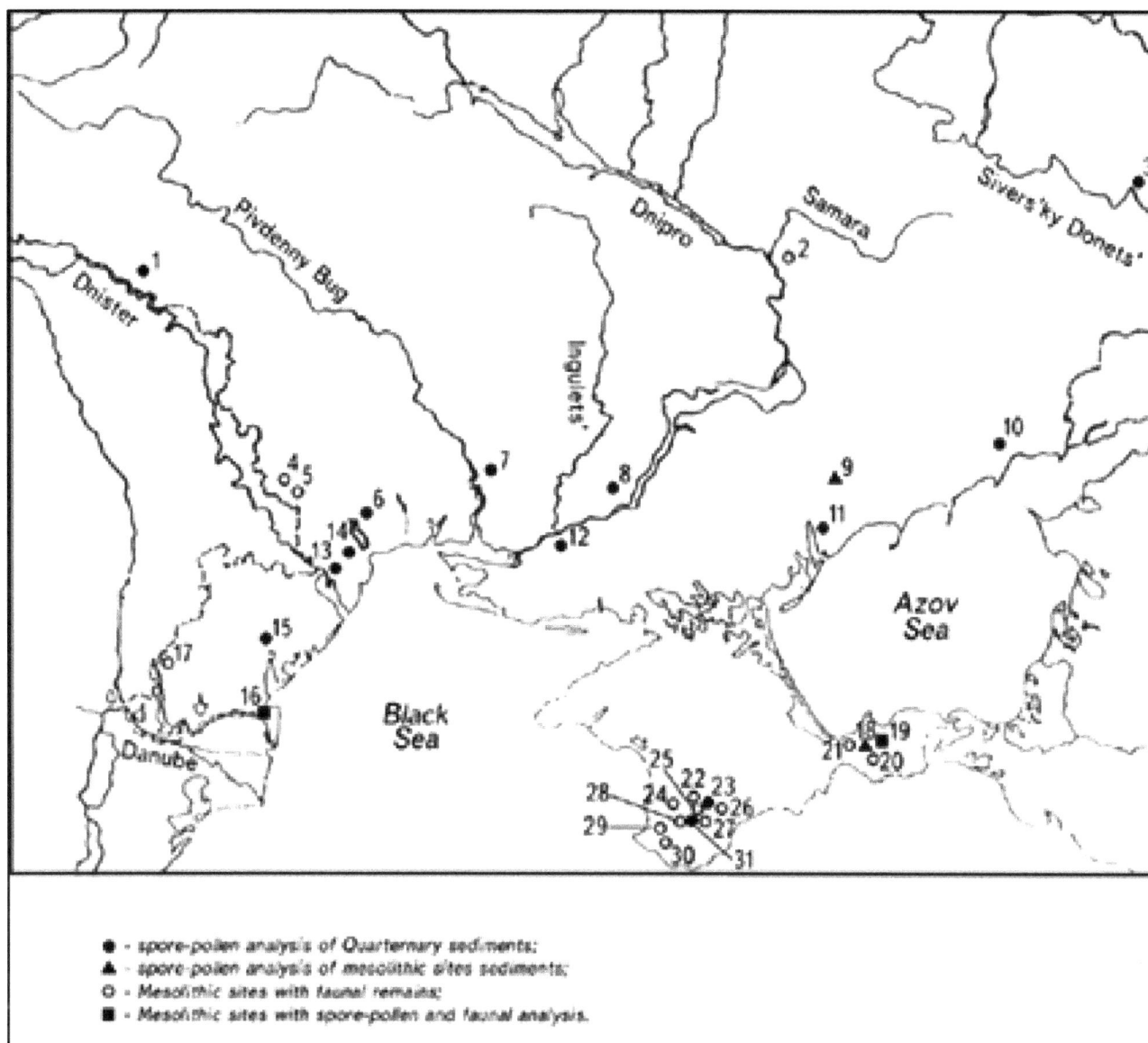

Figure 2. Sources for reconstruction of Boreal geographic situation on the territory of Steppe Ukraine
1 – Murafs'ke, 2 – Igren', 8, 3 – Pidgorivka, Perediel's'k, Rogalyk II, Rogalyk XII, 4 – Girzheve, 5 – Grebenyky, 6 – Kuyal'nyts'ky lyman,
7 – Troïts'ke, 8 – Berislav, 9 – Kam'yana Mohyla, 10 – Myronivka, 11 – Azov, 12 - Kardashyns'ke, 13 – Prymors'ke,
14 – Kryzhanivka, 15 – Prymors'ke (Kiliya district), 16 - Myrne, 17 – Zaliznychne, 18 - Oleksiïvka, 19 – Demerdzhi-Yayla,
20 - Lugove I шар 2, 21 – Lenins'ke I layer 3.

remain the leading arboreal species in the steppe zone [Artyushenko. 1970: 47-51, 60-61, 90-93; Neistadt. 1957: 69-71]. The aurochs *(Bos primigenius Boj.),* prolonging its gradual dispersion to the west, has become the most authentic inhabitant of these semi-closed locales. Most probably, that afforested in such a way plots were mainly associated with flows big and secondary rivers as well as with gulls' springs. In Boreal period of the Holocene the activity of these waterways considerably intensifies in line with general tendency of Black and Mediterranean Sea climate changes.

Therefore, there are serious grounds to consider that in Boreal times total vegetative and animal biomass of Dnister-Dnipro interfluve has increased in comparison with previous period. The enrichment of a food supply contributed to the growth of population density, which was mirrored in sharp increase of settlements' quantity in this region. Nevertheless, the basic

grounds of steppe inhabitants' mode of life were not principally changed. As well as in Early Mesolithic, short time sites represented by extremely small amount of picked up flint artefacts still remaining the leading type of archaeological sites in this region. Only three sites found here (Abusova Balka, Grebeniki and Girzheve) could be conditionally interpreted as seasonal settlements regarding character and quantity of their artefacts. Nevertheless, no diagnostic cultural layer as well as any inner structures or constructive remnants were traced at these settlements.

Such settlements system implies rather high level of mobility of local population, which preferred an extensive way of this living space sources exploitation. Just that's why here one can observe the highest in Steppe Ukraine quantity and density of traces of small collective movements alongside with full absence of long lasting settlements. It rather well

Table 1. Faunal remains at the Steppe Ukraine Late Mesolithic settlements (Boreal period of Holocene).
+ - presence of bones; numerator – MNB; denominator - MNI

Province	Crimea and Azov Steppe				Dnipro Rapids	Lower Danube		Dnister-Dnipro Inter-fluve	
Culture	Girs'ko-kryms'ka		Kukrek Circle of cultures					Grebenikivs'ka	
			Kryms'ka	Dniprov-s'ka	Anetivs'ka				
Site / Fauna species	Frontove I, layer 3	Leninske I, layer 3	Lugove S, layer 2	Kukrek, lower layer	Igren' 8	Mirne	Zaliznichne	Girzheve	Grebeniki
Aurochs (Bos primigenius Boj.)		1/1			?/16	8101/67		19/4	
Bison (Bison sp.)	6/1		9/1						
Bull (Bos sp.)							21/?		
Taurus domesticus	39/5	4/1							
Horse (Equus sp.)	65/7	13/3	1/1		?/2		2/?		+
Horse (Equus gmelini L.)						1369/31			
Horse (Equus caballus L.)								18/3	
Ass (Equus asinus L.)	14/2	4/1	1/1		?/1				
Ass (Asinus hidruntinus L.)	5/1					112/8		10/3	
Wild boar (Sus scrofa L.)				+	?/3	69/6			
Deer (Cervus sp.)	5/3								
Red deer (Cervus elaphus L.)				+	?/12	29/4			
Elk (Alces alces L.)					?/1				
Roe deer (Capreolus capreolus)					?/5				
Saiga (Saiga tatarica L.)						61/6			
Capra-ovis			2/1				2/?		
Ungulata	1/1		1/1						
Wolf (Canis lupus L.)						36/4			
Fox (Vulpes vulpes L.)						4/2			
Lynx (Lynx lynx L.)				+					
Badger (Meles meles L.)						6/3			
Beaver fur (Castor fiber L.)					?/1				
Hare (Lepus sp.)					?/17				
Hare (Lepus europaeus L.)						13/5			
Mammalia						286333			
Aves					+	13/?			
Bustard (Otis tarda L.)						1/1			
Rook (Corvus frugilegus L.)						7/2			
Wild duck (Anas platyrrhyhcha L.)						1/1			
Cormorant (Phalaorocorax carbo L.)						1/1			
Turtle (Emys orbicularis)					+	10/2			
Pisces					?/36	2/?			
Rhutilis frisii Nordm.						2/1			
Pike-perch (Lucioperca lucioperca)					+				
Sheat-fish (Silurus glanis)					+				
Carp (Cyprinus carpio carpio)					+				
Pike (Esox sp.)					+				
Cepea vindobonensis						+			

correlates also with peculiarities of local population hunting activity. The central place in the game structure is taken by small non-gregarious animals, the most fruitful way of hunting on which is individual hunting or hunting by small groups with the help of missile devices with sighting qualities.

Distribution of different ethnic traditions' transmitters in the Dnister-Dnipro interfluve, as it seems now, is also connected with extensive character of this living space exploitation. In fact, the extraordinary situation could be traced in this field: representatives of two different ethnic traditions (Anetivka and Grebeniky cultures) settled here side by side, and in some cases their sites are arranged in immediate proximity. Actually they jointly exploited this living space, not dividing areas of influencing. Moreover, the given locale has become the birthplace of both these traditions, each one of which has local

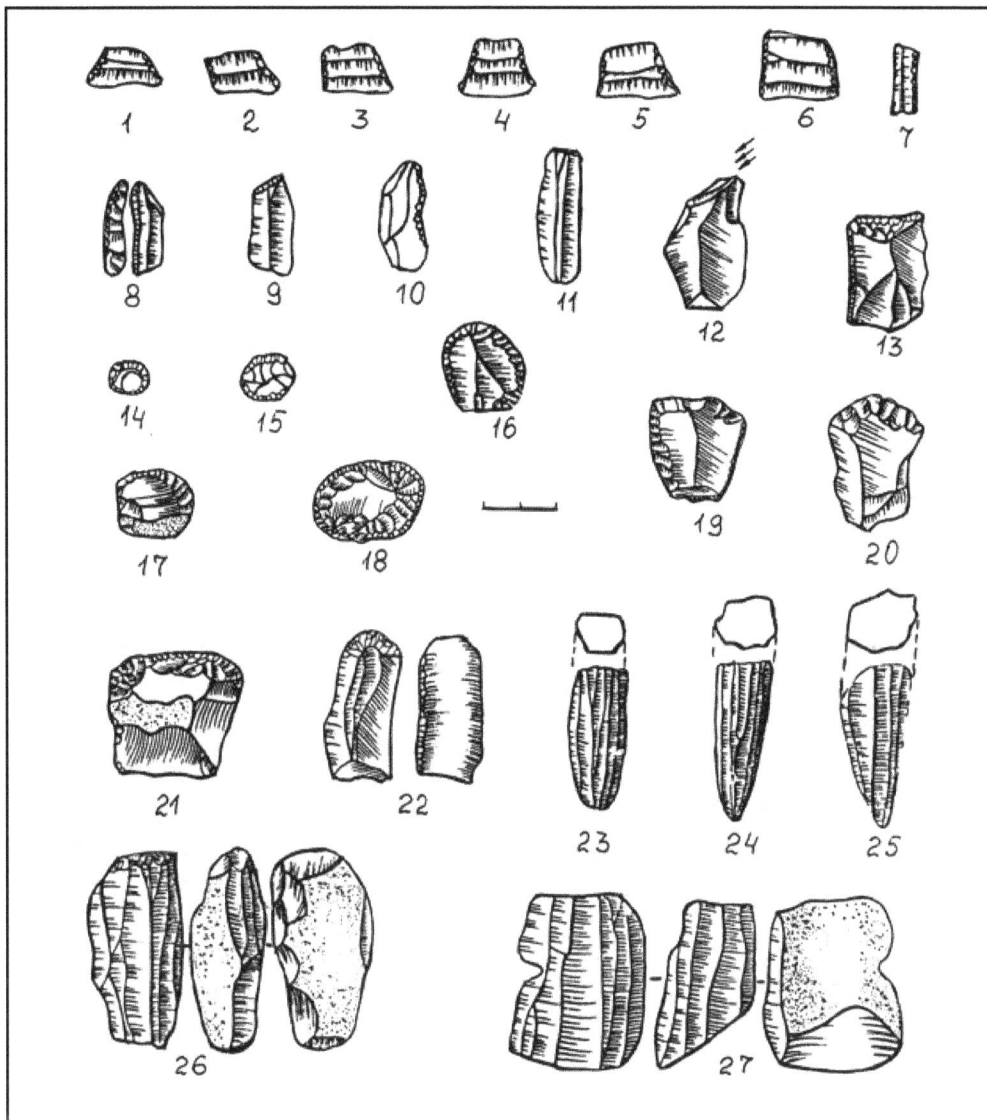

Figure 3. Flint assemblage of Grebeniky culture.

ancestors. Formation of both cultures is the result of gradual evolution of local Early Mesolithic flint processing traditions. Now is it out of doubts that it was the Tsarinka-Rogalik Preboreal cultural circle, which has become the sources for Grebeniki technocomplex origin. In Anetivka Late Mesolithic culture all principal peculiarities of flint artefact morphology inherent to the Preboreal phase of this culture were preserved [Stanko. 1991]. As a result, main bases for difference of their flint assemblages are preserving. Grebeniky primary processing of flint is based on flattened nuclei and characterised by dominance of exact prismatic blades and their cross-sections; in tool complex small circular end-scrapers absolutely prevail, high trapezes are practically the only type of geometrised inserts. Full absence of micropoints and non-geometric microlits is typical for Grebeniky industry (Figure 3). For Anetivka technocomplex, by contrary, diverse types of retouched micro-blades and backed blades inherent; most characteristic forms of tools are blades with ventral processing and burins on massive debitage flakes (Figure 4).

During the last couple of years, attention was paid to these traditions intensive interaction, which actually starts from the moment of their first appearance on historical arena [Kovalenko, Tsoy. 1999: 259]. It should be stressed that traces of such interaction in some extent could be revealed also in the assemblages of other regions of Steppe Ukraine. As it seems now, rather intensive inter-penetration of different in their basic ground culture traditions has become possible due to two circumstances. On the one hand, their ancestors' long lasting life side by side contributed it greatly. On the other hand, the role of Anetivka and Grebeniki population high mobility level could not be underestimated as far as it was just this mode of life, which has caused the realisation of numerous all-level contacts. Their peaceful nature was provided with definite improvement of a food supply. As a whole, irrespectively of a concrete interpretation of the course and consequences of Anetivka and Grebeniky culture transmitters interaction in Dnister-Dnipro interfluve, the fact, that the local population has elaborated rather peculiar understanding of their living space, is out of doubts. Such understanding was based on joined exploitation together with their neighbours since earliest times and has caused improvement of an extensive way of source base utilisation.

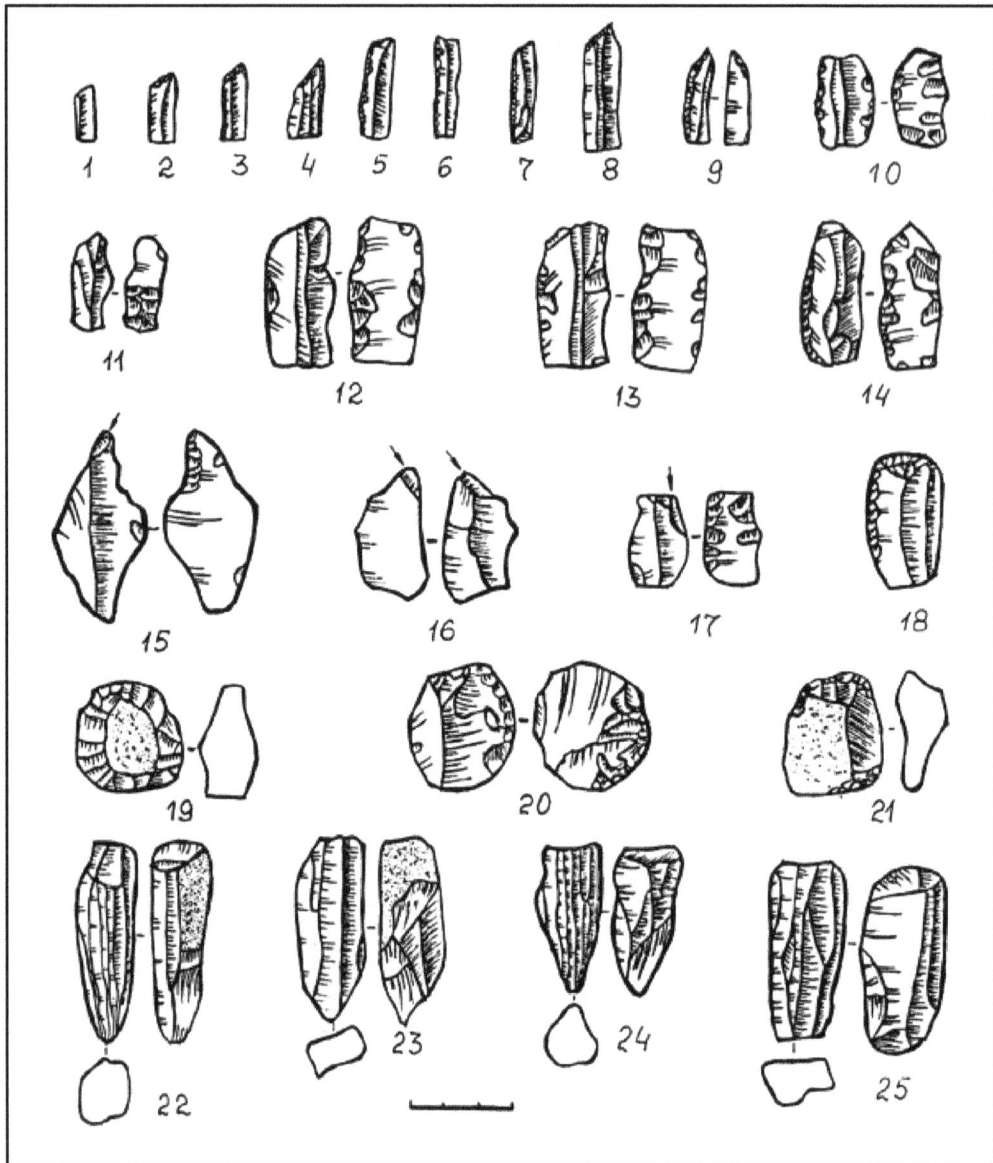

Figure 4. Flint assemblage of Late Mesolithic Anetivka culture.

LOWER DANUBE BASIN

There are series of geographic and cultural peculiarities, which caused delineation of this region as separate living space. On the one hand, it is characterised by most impressive in Steppe Ukraine source base for palaeogeographic situation reconstruction. It is just here absolute majority of faunal remains are concentrated. They give chance to elaborate new understanding of this region inhabitants livelihood system. On the other hand, two biggest Late Mesolithic settlements of Steppe Ukraine investigated during long-lasting excavations are concentrated here. It should be stressed that only one settlement in located strictly in the Lower Danube basin, others are situated mainly in the central part of Danube-Dnister interfluve and are connected with small rivers, ravine springs and with temporary estuaries.

Mesophilous meadow steppe landscape is characteristic for this living space at Boreal times. According to spore-pollen diagram of Mirne, the basic settlement of this region, *Chånopodiàceae* still dominating among grasses, nevertheless the percentage of *Compositae* and *Graminae* is considerably higher in comparison with previous times. Structure of motley grasses has become more diverse and includes *Polygonaceae*, *Fabaceae* and *Rosaceae*. The proportion of arboreal and shrub vegetation also considerably augmented. Compact forest plots are distributed now on declines of rivers and temporary estuaries' plains as well as in girders. In their structure alongside with traditional for steppe zone xerophilous species (e.g. *Pinus sylvestris L.*) deciduous ones - oak (*Quercus sp.*), elm (*Ñarpinus sp.*), linden (*Tilia sp.*), hazel (*Corylus sp.*) and others - are also represented [Pashkevich. 1982]. Afforestation of this region contributed to some rearrangement of faunal species composition. As well as in previous time, central role in it plays the aurochs (*Bos primigenius Boj.*), which prefers semi-closed types of landscape. Besides it, in Boreal period of Holocene diverse forest-steppe and forest inhabitants - wild boar (*Sus scrofa L.*), red deer (*Cervus elaphus L.*), beaver (*Castor fiber L.*) and others - have become important components of Danubian natural environment [Bibikova, 1982: 163]. So, Lower Danube living space seems to be characterised by exceptionally high (if not the highest in Steppe Ukraine) biomass density per square unit. In great measure it was caused by the curious fact that earlier this region practically was not exploited: the only its inhabitants were Bilolissya culture transmitters, lived there at the Dryas III – Preboreal period of Holocene boundary.

Just this richness of Lower Danube basin natural sources seems to be one of the most important grounds of relative stability of local population – Anetivka and Grebeniki culture transmitters. Now it out practically out of doubts, that in the middle – at the second half of Boreal period of Holocene they have come here the territory of their origin is Dnister-Dnipro interfluve. Contemporary source base let us conclude that the newcomers jointly exploited Lower Danube basin. Assemblages of all settlements, known today in this space, contain most characteristic artefacts of both cultural traditions. In some of them (e.g. Mirne) living complexes of Anetivka and Grebeniki population could be rather clearly distinguished in space. In the same time, planigraphic analysis of other sites and settlements give no grounds for such spatial differentiation. Regarding that in Dnister-Bug interfluve analogous situation could be traced in the little bit earlier assemblages, it becomes possible to suppose that Lower Danube living space was inhabited by already integrated Anetivka-Grebeniki population.

Their mode of life in newly exploited territory is characterised by the much greater stability, than in the adjacent spaces. The central base camp of this region's settlers is Mirne, where 18 household complexes connected with broad spectrum of economic functions are discovered [Stanko. 1982]. Most probably, the settlement was inhabited during the whole year. According to standards of palaeoeconomic modelling, no less than 150 persons could live here permanently during at least 9 months. The actual period of this settlement occupation could be more durable as well as there are at about 30 000 non-diagnostic bones, which was not taken into account in the modelling. It seems also highly probable that not all inhabitants of Mirne lived there permanently. A series of synchronous short-term sites arranged nearby the base camp most likely reflect traces of numerous target expeditions organised in order to provide food and raw materials. Apart of the central base camp, there are also at least two seasonal settlements (Zaliznichne and Vasylivka) in the Lower Danube space. Their permanent connection with Mirne seems little probable especially if one takes into account rather significant distance between these settlements. It was rather independent centres, around which life of small collectives was concentrated. In the group short-term sites of the region predominate ones with numerous flint assemblages: quantity of artefact there could be as much as several hundreds; in some cases some kind of flint accumulations could be traced without distinct cultural layer. Such characteristics of short-term sites are not typical for other Late Mesolithic Steppe living spaces.

Specificity of livelihood system of Lower Danube basin also displays itself in the field of subsistence. Faunal remains found here indicate that the most striking features of it are not connected with game structure or with peculiarities of hunting process. As in other areas, basic game species here in Boreal times were aurochs and horse. Considerable increase of small animal proportion confirmed not only by faunal assemblages but also by general dimensions of tools is also typical for the whole Steppe zone. The only hunting peculiarity documented in this region is birds of passage exploitation as game species. The most important feature of Lower Danube basin population subsistence system is connected with elaboration of rather specific attitude to their main hunting species. Significant amount of young aurochs bones together with presence of peculiar knives for grass cutting have created serious grounds to discuss the possibility of this species domestication beginnings. It was supposed that Late Mesolithic hunters have made first steps towards their main game preservation: young aurochs caught at successful hunting was not consumed at once, they were held in special fences, fed up and killed when their meat required [Stanko. 2000]. Two circumstances have contributed greatly to this process. One of them is long lasting utilisation of different bull species by Steppe inhabitants, caused accumulation of deep knowledge about ecology, habits and food base of such game. On the other side, bull (bison) being the subject of special cult of Steppe population was in the centre of their culture system. As a result, new mode of living space food sources management was formed.

Lower Danube basin is the unique region of Steppe Ukraine, where utilisation of gathering products could be proved archaeologically and palinologically.

According to G.A. Pashkevich data, Mirne inhabitants used for food *Chenopodium album, Vicia hirsuta, Polygonum convolvulus*, a dock and other species [Pashkevich. 1982: 136]. Special instruments for vegetative products processing - pestles and graters - were tracologically distinguished in Mirne flint assemblage. Such kind of nutrition was aimed onto carbohydrates and starch proportion levelling, making in such a way subsistence system of Steppe population well balanced. Nevertheless, it should be stressed that seeds and plants gathering had not become the most important food source in Late Mesolithic steppe livelihood. Most researchers believe, that secondary role of this field of economy was connected with specificity of climate and soil, on the one hand, and with local population traditional orientation to the prevailing development of animal breeding techniques [Korobkova.1992: 30].

In general, Lower Danube living space give chance to realise one of the most complicate, stable and versatile system of natural resources exploitation, which we meet in the frameworks of the whole Steppe zone of Ukraine. Possibility of its realisation was guaranteed by relatively high density of vegetational and faunal biomass per unit area. In its turn, such favourable situation was caused by natural course of palaeogeographic process (climate warming and humidifying) as well as by anthropogenious factor (non-exploitation of this region in previous times). First steps of aurochs domestication also contributed it throughout introduction of additional food sources. Diverse and abundant food base improved social relationships and opened perspectives for long lasting peaceful co-existence of different culture traditions.

DNIPRO RAPIDS

Main features of this living space were formed in direct connection with Dnipro River - the principal water flow, which caused some peculiarities of natural environment components as well as determined in some measure livelihood and ethnic culture of this region' inhabitants.

Unfortunately, today there is no direct information concerning the structure of this compact territory floral complex. According to palaeofaunal data semi-closed and closed forest-steppe landscapes absolutely prevailed here. Its typical inhabitants were wild boar *(Sus scrofa L.)*, red deer *(Cervus elaphus L.)*, elk *(Alces alces L.)*, roe deer *(Capreolus capreolus L.)* and beaver fur *(Castor fiber L.)*. On the other hand, the bones of horse *(Equus sp.)* and aurochs *(Bos primigenius Boj.)* indicates presence of the open steppe spaces there.

The most striking features of this living space exploitation could be revealed in the field of local population settlement system. Island settlements are dominating here; nevertheless the general principle of steppe zone settlements' inter-location also is working: there is the central base camp and several less fundamental sites around it. The central place here is Igren' 8, where 10 large semi-deepened huts with inner hearths are situated. Most part of cultural remains (flint and bone artefacts, granite disks, water boiler stones, fauna and plenty of molluscs) are connected with these living structures [Zaliznyak. 1998: 95]. Basing on fauna structure as well as on nature of houses and their organisation the settlement was occupied during cold season. Living objects of this settlement were well equipped for it: isolating layers of cane and molluscs covered their long-pilar skeletons. According to Leonid L. Zaliznyak, these arranged along small river Samara living complexes were not used synchronously; most probably they were formed in the course of repeatedly renovated winter occupation of this territory by the same collective consisted of at about 30 members [Zaliznyak. 1998: 98].

Another specific feature of this living space exploitation is hunter-fishing orientation of local population economy systems. Dnipro islands inhabitants actively explored not only fish, but also water birds, turtles and fluvial molluscs [Nuzhnuy. 1989]. According to Igren' 8 faunal remains, local population preferred big fish, such as pike, soma, pike-perch and carp. Most probably, the mode of its obtaining was closer to hunting than to fishing in modern sense and was realised with the help of sighting missile devices [Neprina. 1988: 29]. It is highly probable that Igren' 8 inhabitants could practise also fishing from boats by plummet nets well known in northern parts of Europe already in Preboreal times

[Zaliznyak. 1998: 98]. Possibility of sporadic usage of boats is indirectly proved by need in transport facilities necessary for connection between islands realisation.

Ethnic structure of this living space is characterised by striking homogeneity, which is not typical for other Steppe zone regions. Dnipro Rapids are included into huge sphere of Kukrek circle of cultures' dispersion, nevertheless peculiarities of flint assemblages found here let us to distinguish separate tradition of this circle – Dnipro one. Its technocomplex is characterised by relatively high percentage of geometric items (in general non-typical for Kukrek tradition), by domination of blade technique as well as by presence of some elements, inherent to cultures of European Boreal Forests (Figure 5). Some influences from the side of Early Mesolithic Dnipro Rapids population as well as from Boreal Sivers'ky Donets' basin inhabitants could be also traced.

One of the most striking specific features of the Dnipro Rapids living space understanding by its inhabitants is connected with archaeological objects, which have no direct analogies in other region of Ukraine: that is cemeteries. For the first time they appear here already in Allerod, and since that time funerary rite and cemetery organisation system has been seriously modified. Nevertheless, the basis of mental value of objects of this genre in local population ideology was not changed. Most researchers believe that presence of such types of cultural objects is connected with existence of continuos and stable links of local population with its living space and with special sort of its mental reflection. It could not be excluded that the notion of "Motherland" or "native territory" notion could be elaborated and realised just in such context.

CRIMEA AND AZOV STEPPE

Delineation of this territory into separate living space is caused not so much by its natural geographic environment peculiarities as by specificity of its inhabitants' ethnic structure. As well as in Dnister-Dnipro interfluve, available now source base concerning northern part of this region does not allow to reveal vegetation character and species composition substantial shifts, which could be connected with transition to Boreal period of Holocene. Only on Kerch peninsula spore-pollen and faunal complexes could be chronologically attributed rather exactly.

According to such data, open steppe landscape prevailed in the whole Crimea-Azov living space. In its vegetation complex species well adapted to arid and sharply continental climate *(Chenopodiaceae* and *Artemisia)* prevailed. Proportion of arboreal species in the majority of diagrams does not exceed 7-10 %, in their structure the representatives of steppe flora - pine *(Pinus sylvestris)*, birch *(Betula nana)* and other xerophilous species dominated. Horse *(Equus sp.)* and ass *(Equus asinus L.)* are typical inhabitants of such arid space. At the same time, as well as in other steppe regions, local palynological complexes testify also presence of forest plots with deciduous species participation. In Dnipro-Azov

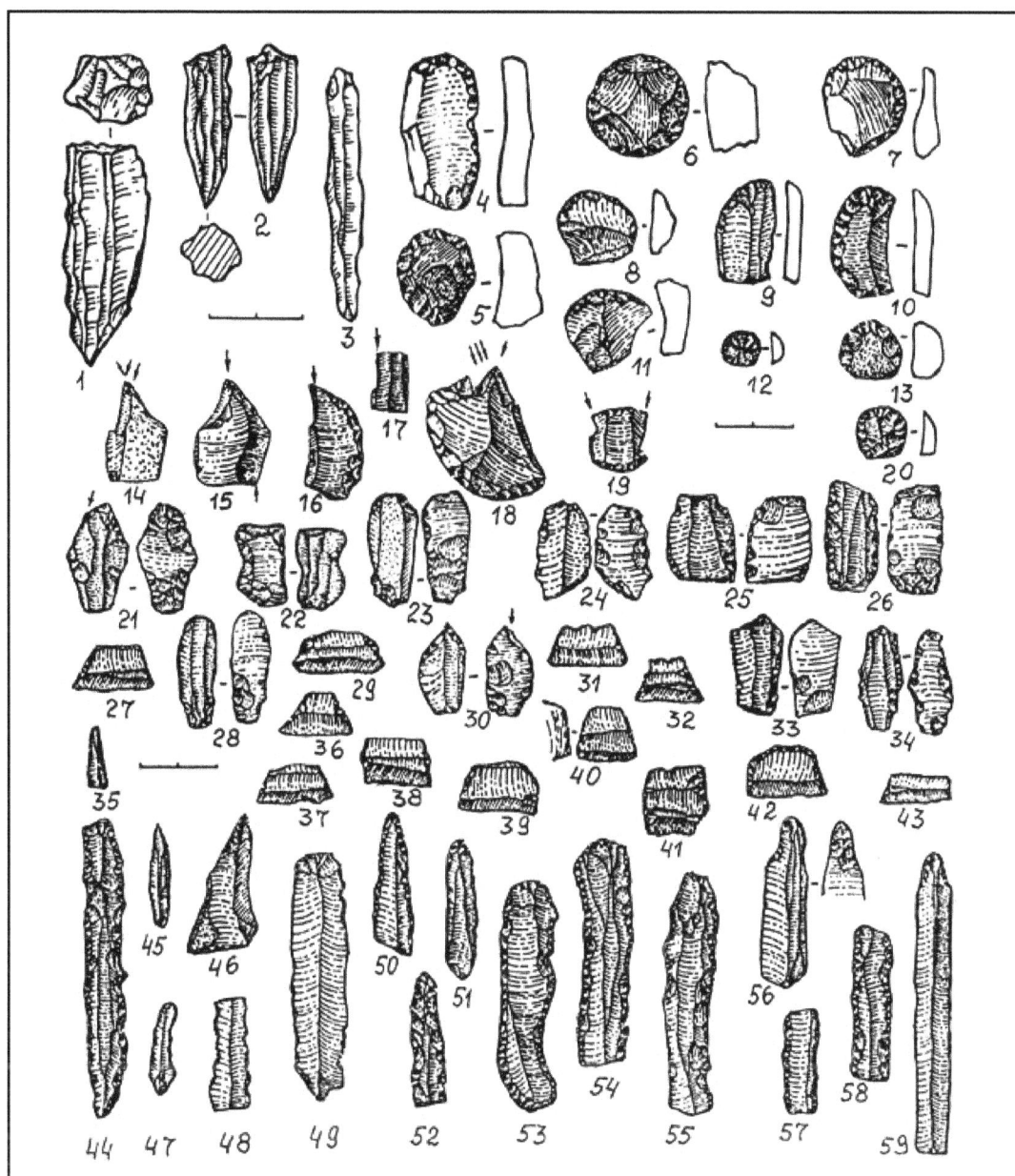

Figure 5. Flint assemblage of Dnipro Kukrek culture [Nyzhnyi, 1989].

steppes among the latter most widespread is alder *(Alnus sp.)* and in Kerch peninsula - linden *(Tilia sp.)* [Matskevoy, Pashkevich. 1973]. Most probably, these narrow forest plots, which were connected mainly with river basins and alternated there with motley grass segments, have become places of aurochs and deer of unknown species distribution in the Kerch peninsula. General density of vegetative and animal biomass in Crimea-Azov Steppe region was of one of the lowest within the frameworks of Steppe Ukraine. Such characteristics of potential food base have caused some peculiarities in this region exploitation.

As in other steppe regions, large long lasting base camps are absent here; the majority of sites are attributed as places short-term occupation, traces of which are represented exceptionally by collected on the surface small number of flint artefacts. Only one settlement (Frontove II, layer 3) could be interpreted as seasonal camp of simplified mode. Remnants

of surface dwelling with stone foundation were opened there; near the entrance to it flint artefact and raw material store was traces [Matskevoy. 1977: 72, 131]. This construction was used, most probably, for only one season, but the cultural layer accumulation took place during several years. Flint assemblages structure of Kam'yana Mohyla and Velyki Kopani settlements also does not exclude the possibility of their utilisation as seasonal camps. As a result the population density per square unit in the Crimea-Azov Steppes seems to be the lowest one in the frameworks of the whole steppe Ukraine.

Rather severe palaeogeographic conditions and, first of all, extremely low density of vegetative and animal biomass per unit area acted also as local population ethnic structure regulators. So, in particular, the Anetivka culture transmitters' preadaptation level did not allow them to move into the Crimea-Azov Steppe space further than Lower Dnipro basin.

Sources of this space were unsatisfactory also for Late Mesolithic Donets'ka culture, transmitters of which tried to explore only northern part of the region under study.

Basic ethnic unit here was Crimean culture of Kukrek culture circle, which was formed just here, in the latest phases of Preboreal period of Holocene. Peculiar pencil-shaped nuclei, "Kukrek" burins on massive flakes and debris as well as absolute dominance of non-geometric forms in microlithic inserts complex are characteristic for flint industry of thus culture during all the period of its existence. Most striking type of artefacts are blades or their medial sections with ventral processing, called "Kukrek inserts"(Figure 6). At Boreal stage of this culture development its transmitters successfully explore not only Azov Steppes but piedmonts and pastures of Crimea Mountains and Kerch peninsula as well. At the end of Boreal times this culture traces could be revealed also in assemblages of inner Crimea Mountains' settlements. So, it becomes possible to suppose that this

dissemination of Crimea Kukrek population reflects the processes of Crimea and Girs'ko-kryms'ka cultures interaction results of which could be seen in Kerch peninsula ethnic map.

So, analysis of Crimea-Azov Steppe living space let us to distinguish two sub-regions which are characterised by different feature of their palaeogeography as well as by some peculiarities of local population ethnic structure. Kerch peninsula, one of such sub-regions, was jointly exploited by Crimea Kukrek and Girs'ko-kryms'ka culture transmitter. It was characterised by relatively higher exploitation intensity rate in comparison with the second sub-region of this living space - Azov Steppe one. Difference in biomass density per unit area seems to be the basic reason for such distinction. On the other hand, available now source base give no serious grounds to argue that revealed here variability is the same nature as one, traced among different living spaces of the Steppe Ukraine.

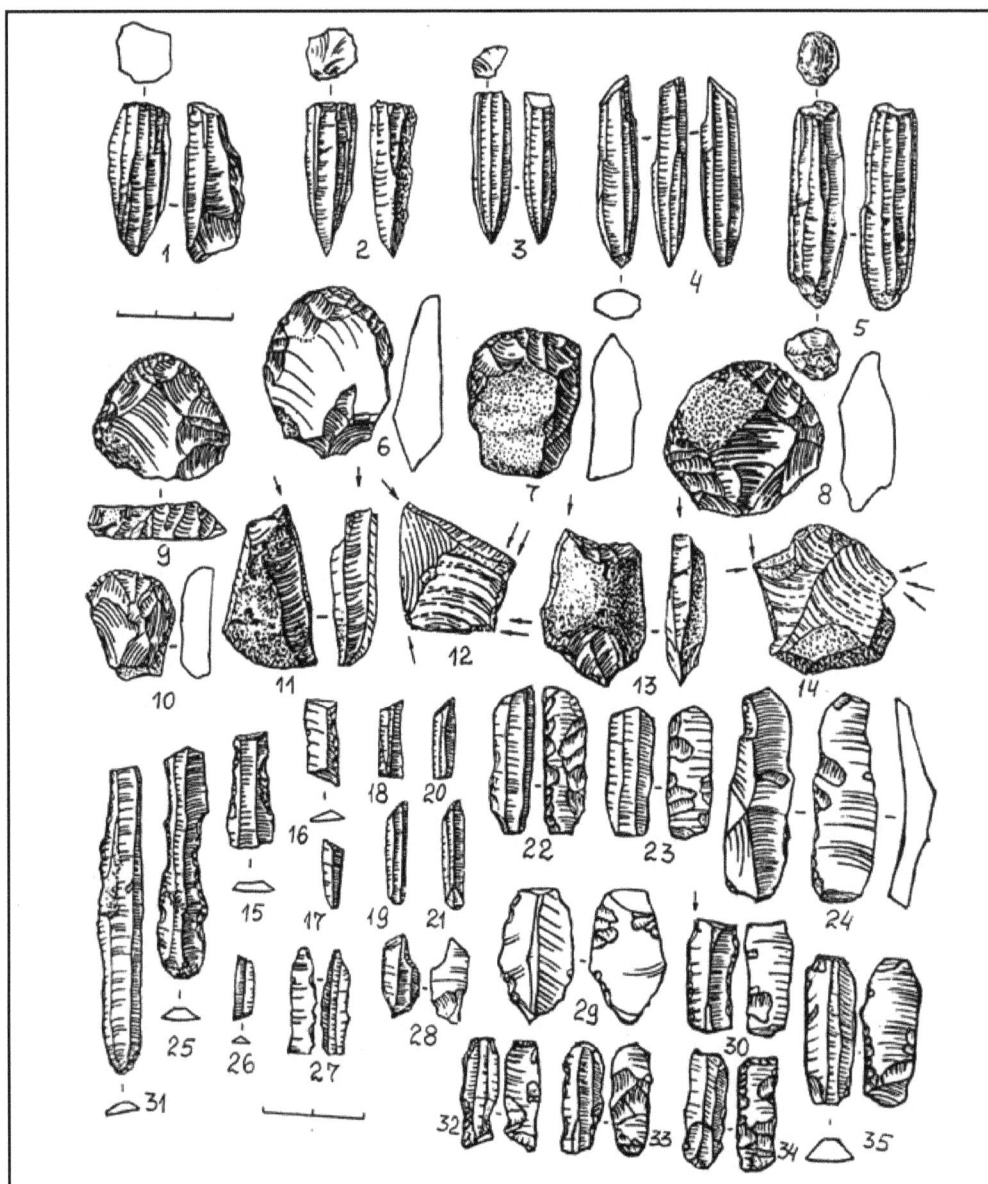

Figure 6. Flint assemblage of Crimea Kukrek culture.

SIVERS'KY DONETS' BASIN

This space was characterised by series of palaogeographic peculiarities as well as by some specific features of their exploitation during the whole Mesolithic period. In Boreal period of Holocene natural geographic and cultural heterogeneity inherent to this region in previous time displays itself but in smaller extent.

At the beginning of Boreal period of Holocene steppe landscape was absolutely dominated at the upper parts of the Syvers'ky Donets' basin. In their spore-pollen diagrams percentage of mesophilous motley grasses is as much as 60 %; among them *Rosacea, Lamiumae, Astraea* and *Fabales* prevail. In the group of arboreal species pine (*Pinus sylvestris*) dominate, but pollen of deciduous species, in particular, of elm (*Ulmus sp.*), also presents [Gerasimenko. 1996: 24]. On the lower parts of the region, on watersheds' declines and in girders mixed forest-steppe landscape prevailed. In their spore-pollen spectrums percentage of xerophilous vegetation does not exceed 7 %, and in spite of predominance of birch *(Betula nana)* in composition of their arboreal species, the proportion of oak *(Quercus sp.)* is as much as 18 %. Microfossils of *Alnus sp., Malus sp.* and *Lonicerae* are also typical for such motley grasses' plots. At the end of Boreal period of Holocene some kind of vegetation xerophitisation could be traced here: proportion of motley grasses in all parts of this space decreases, and *Ephedrae* and *Chenopodiae* become the most widespread species in this region [Gerasimenko. 1996: 24].

In contrast to natural geographic environment, local population living space exploitation systems are rather homogeneous one. Syvers'ky Donets' basin is considered as one of the most consolidated in ethnic sense regions of Steppe Ukraine. The principal ethnic unit here was Donets'k culture, sites of which could be divided into two groups. Technocomplex of its earlier, Teplivs'ka, group is characterised by presence of very small circular scraper, close to Grebeniky ones, as well as by series of low and not very

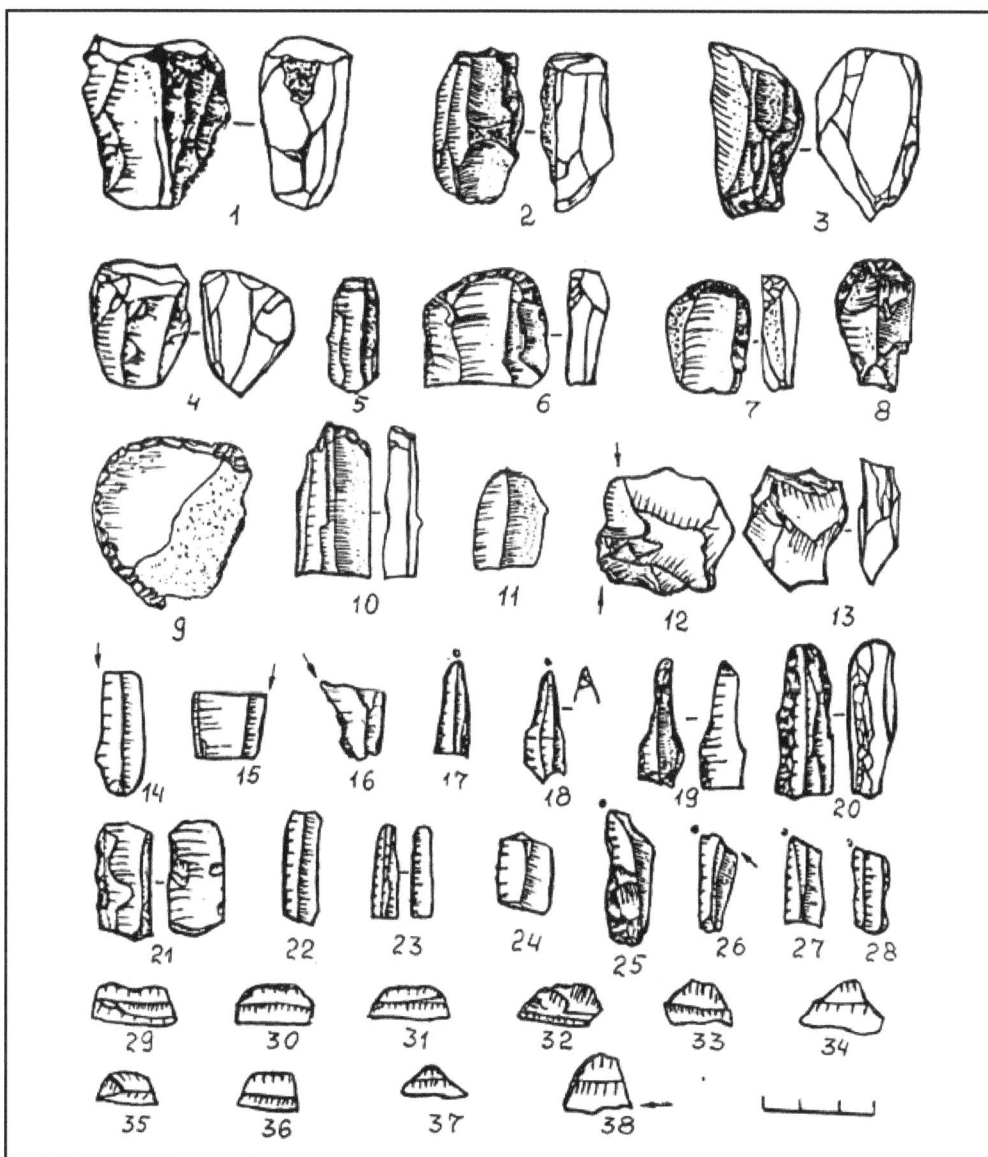

Figure 7A. Flint assemblage of Donetsk culture: Teplivs'ka group [Gorelik, 1997].

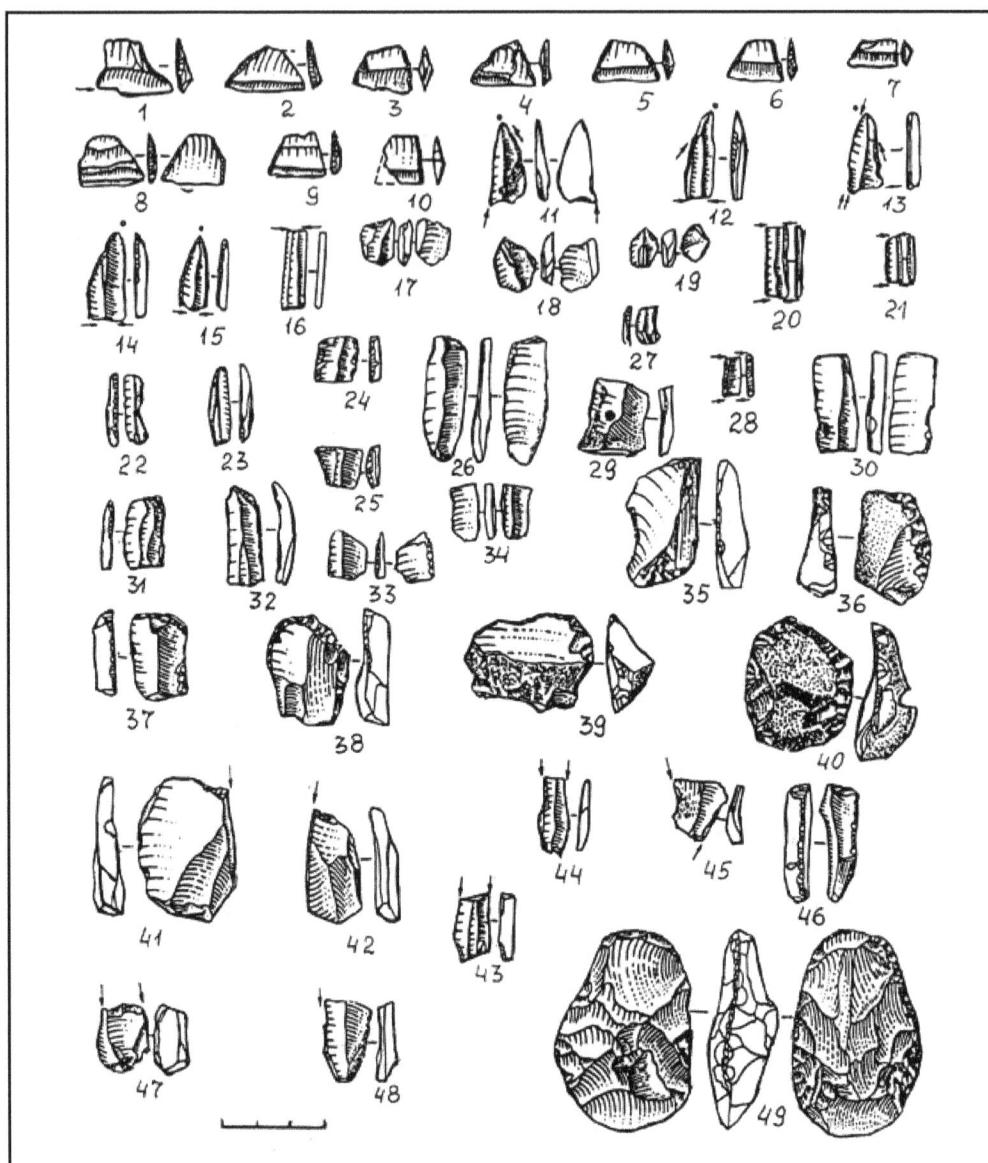

Figure 7b. Flint assemblage of Donetsk culture: Pelageivs'ka group [Gorelik, 1997].

high trapezes. In assemblages of the later, Pelageivka group, elements typical for so called "Forest" cultures (Janislavice points and micro-burins, asymmetric triangles and oval hatches etc.) could be traced (Figure 7: A, B). Today most researchers interpret Donets'ka culture as heterogeneous phenomenon, in formation of which role of Steppe and Forest inhabitants' migrations could not be underestimated [Gorelik. 1997; Zaliznyak. 1998]. The necessary background for such for this process has been created by the diversity of natural geographic palaeoenvironment, where components of different landscapes co-existed side by side. The importance of Syvers'ky Donets' palaeogeography for such cultural phenomenon development is also proved by the fact, that Donets'ka culture transmitters were distributed only in this space, not going far away from this big river.

Their settlement system in its basic features is analogous to one, known in other living spaces of Steppe Ukraine. There are the central place (Shevchenko) and many short-term sites around it. Nevertheless, in contrast to Dnipro and Danube

basins, it is impossible to reveal distinct site "clusters" here. Character of central settlements also differs in some extent: it is hardly possible to interpret it as base camp, rather it looks like seasonal camp without any constructions. Extremely rich flint source base together with general stabilisation of life has given chance to organise special sites of flint primary processing (Petropovlivka). Such type of sites existed here at Palaeolithic times and unknown here for Early Mesolithic period.

In the central part of Sivers'ky Donets' living space there are some traces of Early Mesolithic tradition of Mynivskiy Yar gradual evolution, and assemblages of Rubtsi type have become the result of it. Characteristic for them presence of peculiar axes, post-Sviderian points and by full absence of geometrical inserts, as many researchers believe, create serious grounds to make some references with culture of Ukrainian Polissya [Gorelik, Dukhin. 1984]. Unfortunately, today only two sites represent this tradition, and that's why we should be very careful when interpreting it.

It is difficult enough to reconstruct also subsistence system of Syvers'ky Donets' inhabitants as far as there is no faunal remains at the sites. Basing on general characteristic features of this region palaeogeography, it becomes possible to suggest that systematic hunting for fish, for water birds and fluvial molluscs collecting were rather important part of local hunter-gatherers [Gorelik. 1997].

In general, peculiarities of Syvers'ky Donets' region takes as separate living space could be traced in the field of this region palaeogeography as well as in the livelihood systems, ethnic structure and cultural links of its inhabitants

CONCLUSION

So, brief analysis of specific features of five living spaces distinguished in the frameworks of Steppe Ukraine let us outline some aspects where differences of these spaces display themselves rather distinctly. First of all, peculiarities could be revealed in main components of palaeogeographic environment, in vegetation and fauna species structure. Basic elements of local population livelihood systems - settlement' structures, sites systems, dwellings, subsistence activity etc. - are also different. It should be also stressed substantial difference revealed in the result of ethnic interpretation of most of living spaces. It is very important to pay special attention to the very interesting collision: in some cases living spaces differ according some of criteria, seem to be analogous according to other one. It is possible to suggest that such affinity of some parameters of living space exploitation systems could be caused by two basic factors. On the one hand, general tendency of Steppe palaoegeography and culture evolution could be reflected here. On the other hand, this affinity might be the result of Kukrek tradition transmitters spreading into the most part of the territory under study (except Syvers'ky Donets' basin). So, ethnic implication of living space definition and development of its exploitation system becomes the issue of principal importance. It becomes especially acute in the context of general problem of correlation of ethnic and ecological background in adaptation systems of Early Prehistoric population.

Author's address

Dr Olena V. SMYNTYNA
Head of Department of Archaeology and Ethnology of Ukraine
Odessa I.I. Mechnikov National University
2, Dvoryanska str.
Odessa, 65026 UKRAINE

Bibliography

ARTYUSHENKO, A.T., 1970, *Rastitelnost' Lesostepi i Stepi Ukrainy v Cetvertichnom Periode.* Kiev: Naukova dumka.

BIBIKOVA, V.I., 1982. Teriofauna poseleniya Mirnoye. In STANKO, V.N., 1982, p. 139-164.

GERASIMENKO, N.P., 1996, Prorodnaya sreda obitaniya cheloveka na yugo-vostoke Ukrainy v pozdnelednikoviye i golocene. *Archeologicheskiy Almanakh* 6, p. 3-64.

GORELIK, A.F., 1997, Slozheniye donetskoy kultury i nerotorye metodologicheskiye problemy "neolitizatsii" mesoliticheskikh kultur. In *Archeologia i ethnologia Vostochnoy Evropy: materialy i isslodovaniya,* edited by S.A. Bulatovich. Odessa: Hermes, p. 123-132.

GORELIK, A.F. & DUKHIN, A.O. 1984. Raskopki mesoliticheskikh pamyatnikov Minievskiy Yar i Rubtsy. In *Novye archeologicheskiye issledovaniya na Odeschine,* edited by G.A. Dzis-Rayko. Kiev: Naukova dumka, p. 13-24.

KOROBKOVA, G.F., 1992, Istoki neolitizatsii: k probleme khozyaistvenno-kulturnogo razvitiya Severo-Zapadnogo Prichernomoria v epohu mesolita. *Studia Praehistorica* 11-12, p. 28-34.

KOVALENKO, S.I., TSOY, V.B., 1999, K voprosu o razvitii pozdnemesoliticheskikh industriy v Karpato-Dunaiskom regione. *Stratum Plus* 1, p. 257-262.

MATSKEVOY, L.G., 1977, *Mesolit i Neolit Vostochnogo Kryma.* Kiev: Naukova dumka.

MATSKEVOY, L.G.& PASHKEVICH, G.A., 1973, K paleogeografii Kerchenskogo poluostrova vremen mesolita i neolita. *Sovetskaya Arheologia* 2, p. 123-137.

NEISHTADT, M.I. 1957. *Istoriya Lesov i Paleogeographiya SSSR v Golocene.* Moskva: Izdatelstvo Akademii Nauk SSSR.

NEPRINA, V.I., 1988, Vynyknennya to rozvytok rubal'stva na territorii Ukrainy. *Arheologia* 64, p. 28-33.

NUZHNYI, D.Yu., 1989, O svoeobrazii pamyatnikov Kukrekskoy kulturnoy traditsii v Dneprobskom Nadporozhie. In *Kamennyi vek: pamyatniki, metodika, problemy,* edited by S.V. Smirnov. Kiev: Naukova dumka, p. 145-154.

PASHKEVICH, G.A., 1982. Paleobotanicheskaya characteristika poseleniya Mirnoye. In STANKO, V.N., 1982, p. 132-138.

SMYNTYNA, O.V., 1999, K probleme tipologii mesoliticheskih pamyatnikov. *Stratum Plus* 1, p. 239-256.

STANKO, V.N., 1982, *Mirnoye. Problemy Mesolita Stepey Severnogo Prichernomoriya.* Kiev: Naukova dumka.

STANKO, V.N., 1991, Kulturno-istoricheskiy process v mesolite Severo-Zapadnogo Prichernomoriya. In *Severo-Zapadnoye Prichernomoriye: kontaktnaya zona drevnikh kultur,* edited by V.P. Vanchugov. Kiev: Naukova dumka, p. 5-17.

STANKO, V.N., 2000, Pervye skotovody azovo-prichernomorskih stepey. In *Archeologia ta ethnologia Skhidnoyi Evropy: materialy i doslidzhennya,* edited by O.V. Smyntyna. Odessa: Astroprint, p. 7-20.

ZALIZNYAK, L.L., 1998, *Peredistoria Ukrainy Õ-V tys. do n.e.* Kiev: Biblioteka ukraintsya.

SCYTHIA BEFORE THE SCYTHIANS

P.M. DOLUKHANOV, M.L. SÉFÉRIADÈS & V.N. STANKO

INTRODUCTION

The project under this title was carried out in 1998 and 1999 in the north-west Black Sea area, which corresponds to the western part of the Odessa Oblast (Province), Ukraine. The international research team included Dr. Dolukhanov (Co-ordinator); Professor Stanko (Principle Investigator) and Dr. Séfériadès (Co-Director). Field excursions were carried out in August 1998 and in August 1999. In these excursions took part Dr G. Pilipenko (Department of Geography at the Odessa State University), Dr E. Smyntyna (Department of Archaeology and Ethnology at the State University of Odessa), Dr. L.V. Subbotin (Museum of Archaeology, Odessa) was responsible for the excavations and survey of Gumelnitsa sites in August 1999. Anne Dambricourt-Malassé (Institute de Paléntologie Humaine, Paris) carried out the study of the skull fragment that had been found during the excavations.

The main objectives of the project were:

1. Reconstruction of the Mesolithic and Chalcolithic landscape;

2. Assessment of the spatial patterning, social dynamics and the use of resources by Mesolithic hunter-gatherers and Chalcolithic farmers.

Prehistoric sites in this area have been intensively studied over the past 50 years. One should especially mention a large-scale archaeological project of the 1950-60s: the Danube-Dniestr Archaeological Mission directed by N. M. Shmaglii. The survey and excavations were also conducted by S. N. Bibikov, P.I. Boriskovsky, V. N. Stanko, L.V. Subbotin and many other archaeologists. These studies resulted in the discovery and excavations of a large number of prehistoric sites, ranging in age, size and character, which were reported in several major publications.

Basing on the existing evidence and keeping in mind the above-stated objectives, we carried out the following investigations:

a) Detailed landscape survey in the key areas of archaeological sites;

b) Archaeological reconnaissance and the establishment of the exact location of archaeological deposits in relation to landscape features with the use of GPS equipment;

c) Test excavations aimed at establishment the exact stratigraphy and taking the samples for radiocarbon dating, pollen and soil analyses.

NATURAL ENVIRONMENT

The area under investigation is located in the Danube-Dniestr interfluve, which forms the western segment of the North-Pontic Lowland (the historical Scythia), a large hollow in the crystalline basement of the East European Plate. Its surface was formed by an undulating loess-covered plain and slightly tilted to the south, in the direction the Black Sea. The plain is crossed by several rivers; both the large (the Dniestr, Kogulnik, Sarata), and the small (the Dracula, Nerushai and many others). The lower stretches of the larger rivers form often impressive estuaries (or the *limans*).

During the Last Ice Age, due to the global eustatic regression of the sea-level, the Black Sea was separated from the Mediterranean basin and became a large land-locked lake with the level at 90 ÷ 100 m below the present one (Aksu et al, 1999). The coastal vegetation consisted of xerotic grassland with *Chenopodiaceae* and *Artemisia* (Artyushenko 1970).

During the Late Glacial period,12,000 and 9,500 BP, as the climate grew milder and wetter, an intensive melting of glaciers led to the rapid rise of the Black Sea level. Between 7,200 and 7,000 years BP, the Black Sea re-established the links with the Mediterranean Sea via the Bosphorus. As a result, sea water ingressed into the river valleys and formed the estuaries which gradually developed into the present-day limans (Fedorov 1978; Aksu et al, 1999).

The pollen analysis shows that rare forests started spreading along the river valleys from c. 9,000 years BP onwards. Initially they consisted of pine and birch, and later included the broad-leaved species: oak, elm, and the understory of hazel-nut; they were restricted to the bottomland and terraced slopes. The steppe vegetation on the watersheds gradually acquired a mesophytic character. Precipitation in the entire Mediterranean area markedly increased 8 - 7 thousand yr. ago (Rohling et al. 1999). In the Pontic area this period coincided with the maximum spread of mixed coniferous-deciduous forests in the uphill and the valley bottomland. The silvo-steppe with isolated stands of oak, elm, lime and maples extended over the watershed plain (Kremenetsky 1991).

MESOLITHIC

So far no radiocarbon dates are available for the Mesolithic sites in the studied area. Yet, judging from the dates obtained in the neighbouring regions of Ukraine and Moldavia, the

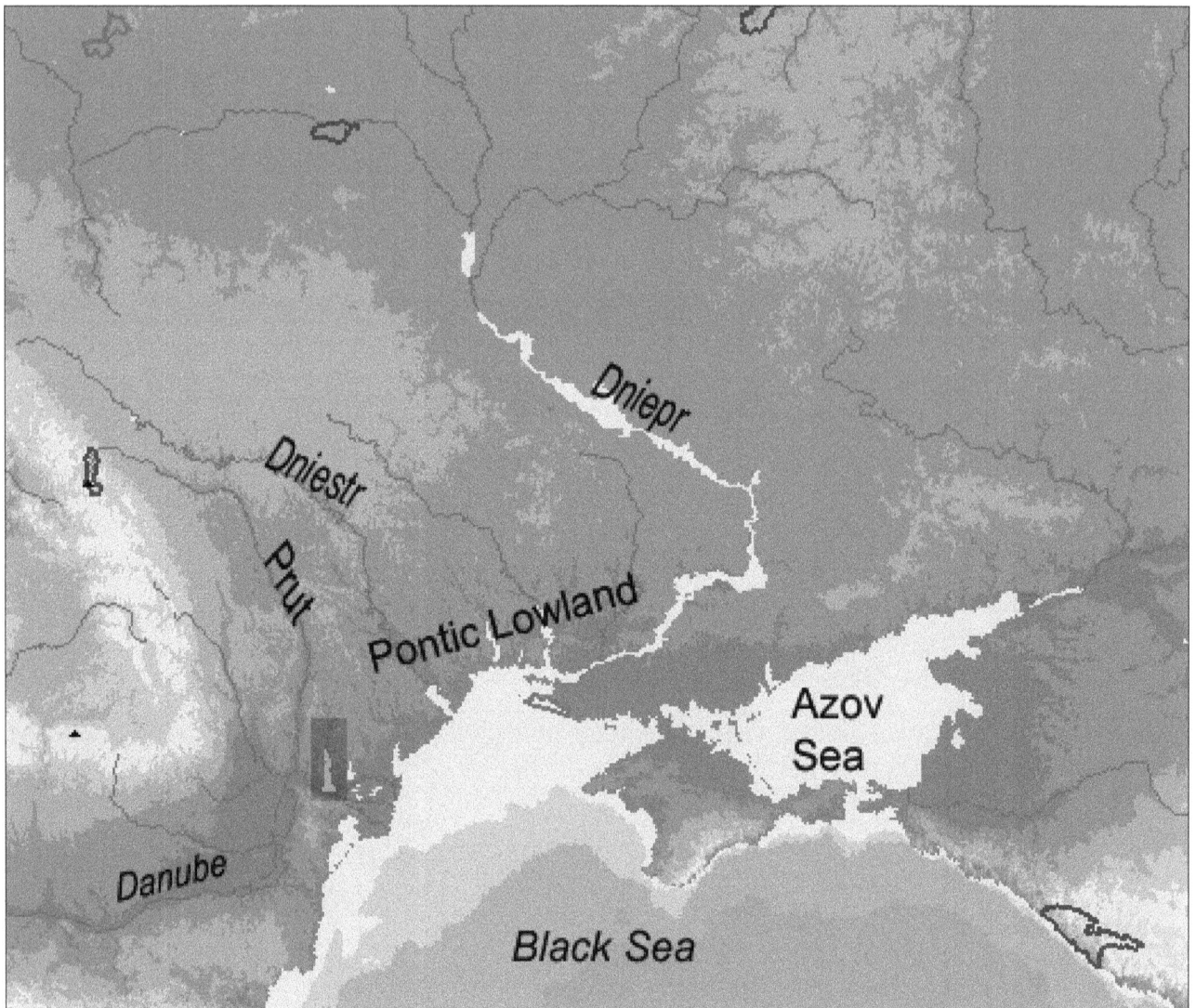

Figure 1. General situation.

age of Mesolithic may be assessed as 10,000 - 7,500 bp (or 8,200 - 5,400 years BC cal.).

The largest site, Mirnoe 1, was located on the floodplain of the Dracula river, 800 m to the south-west from the southern limit of the village of Mirnoe. The site was discovered in 1963; systematic excavations were carried out in 1969-1976 by V.N. Stanko, resulting in the exposure of 1807 sq. m of archaeological deposits (Stanko 1982). A 'hearth area' was acknowledgeable in the central part of the settlement, it included several hearths and so-called 'backers' pits', with a high concentration of animal bones and flint pieces. As suggests the use-wear analysis, lithic tools were used predominantly for butchering and skin dressing. Retouched blades are also numerous, while the cores are few in number. 18 clusters of implements were identified in the peripheral part the site, their boundaries not always being clear. Among 214 animal bones found at the site were identified: horse (minimum number of individuals - 7); aurochs (3), saiga (3) and fox (1). The finds of elongated prismatic blades with characteristic 'sickle-gloss' were particularly significant; basing on the use-wear analysis they were identified as 'reaping knives' intended for harvesting edible plants.

Several potentially edible plants were identified by Pashkevich in the cultural deposits (1982): white goosefut (*Chenopodium album*); black bindweed (*Polygonum convolvules*); hairy vetch *(Vica hirsuta)* and sorrel (*Rumex acetosa).*

The second largest site, Beloles'e, was also located on a low-lying level: the flood-plain of the Sarata river, 2 km upstream from the point where the river empties into the Sasyk liman. The site was excavated by V.N. Stanko in 1965-1966 and 1977. The total area of excavated archaeological deposits was 180 sq. m. Four structural units with hearths have been identified within this area. Blades were most commonly used for manufacturing the implements with secondary trimming. Scrapers formed the largest group which included several varieties: the circular end-scrapers on blades and flakes, scrapers on flake, circular and sub-circular on flakes and others. The lithic inventory included burins, points, backed bladelets and geometric microliths: crescents, trapezes and rectangles. Animal remains consisted of wild horse (57.1%); aurochs (28.1%) and saiga (14.3%).

Two Mesolithic sites: Trapovka and Borisovka were found in identical positions, respectively, along the western and eastern bank of the Sasyk liman, immediately above the abrasion cliffs. At the time of the existence of the sites, the sea-level was considerably lower and the both sites were facing the flood-plain of a single river. Among the implements were identified: four prismatic and one *bonnet de mître*-type cores; end-scrapers on blades; Kukrek-type blades; several fragments of trapezes, micro-points and micro-blades.

The site of Vasilievka was located also on the fringes of the plain facing a deep slope of the eastern bank of the Kitai *liman*, at the altitude of 20 m above sea-level. The general character of the industry is microlithic, with the common occurrence of trapezes and Kukrek-type blades.

The site of Zaliznichnoe was unique by its topographical position: high on the slope of the valley of the Yalpug River, 125 - 130 m above sea-level. The site was excavated during the 1999 field project in an area of 24 sq. meters (Fig. 2). The artefacts and animal bones (not yet identified) were found in the loamy silt directly beneath the soil, at the depth of 0.10 - 0.55 cm. The collection of lithics includes c. 2,000 pieces. The industry is of a microlithic character (Fig. 3). Trapezes, truncated points, Kukrek-type blades and large points were the most common in the inventory, which combines both Mesolithic typological traditions that had been earlier established in the greater area of Ukraine: Grebenikian and Kukrekian (Stanko 1982).

Summing up our preliminary observations one may conclude, that the largest sites in the studied area (Mirnoe 1 and Beloles'e) lie on the lowermost levels. These sites which supposedly included several blood-related families may be viewed as base-camps inhabited on the round-the-year basis. Trapovka and Borisovka were allegedly winter sites of limited duration. The sites of Vasilievka and Zaliznichnoe were located at the higher altitude, in close proximity to the watershed surface. Zaliznichnoe was also located at the greater distance from the main water channel. A suggestion is made that that these sites, like other sites found in the similar setting further to the north, resulted from seasonal migrations (the summer transhumance) of smaller groups stemming from the base- camps

GUMELNITSA

The study of Gumelnitsa Culture started in 1924 when the site of that name was discovered in Romania, on the left bank of the River Danube (Dumitrescu 1924). The first Gumelnitsa site to be found in the Odessa Oblast, was Bolgrad discovered by I.T. Chernyakov in 1960 in the town of that name. The sounding of this site started the next year and the systematic excavations followed since 1962. To this day 31 sites of Gumelnitsa Culture were found in the Odessa Oblast and the neighbouring Moldavia. These sites corresponded to the earliest farming communities to be recognised in this entire area. Their age may be estimated basing on two radiocarbon dates obtained for the site of Vulcanesti III in the southern Moldavia: MO-417: 5810 ± 150 years bp (4896-4503 years BC cal.) and LE-640: 5300 ± 60 years bp (4235-4006 years BC cal.). Animal remains from the Gumelnitsa sites have

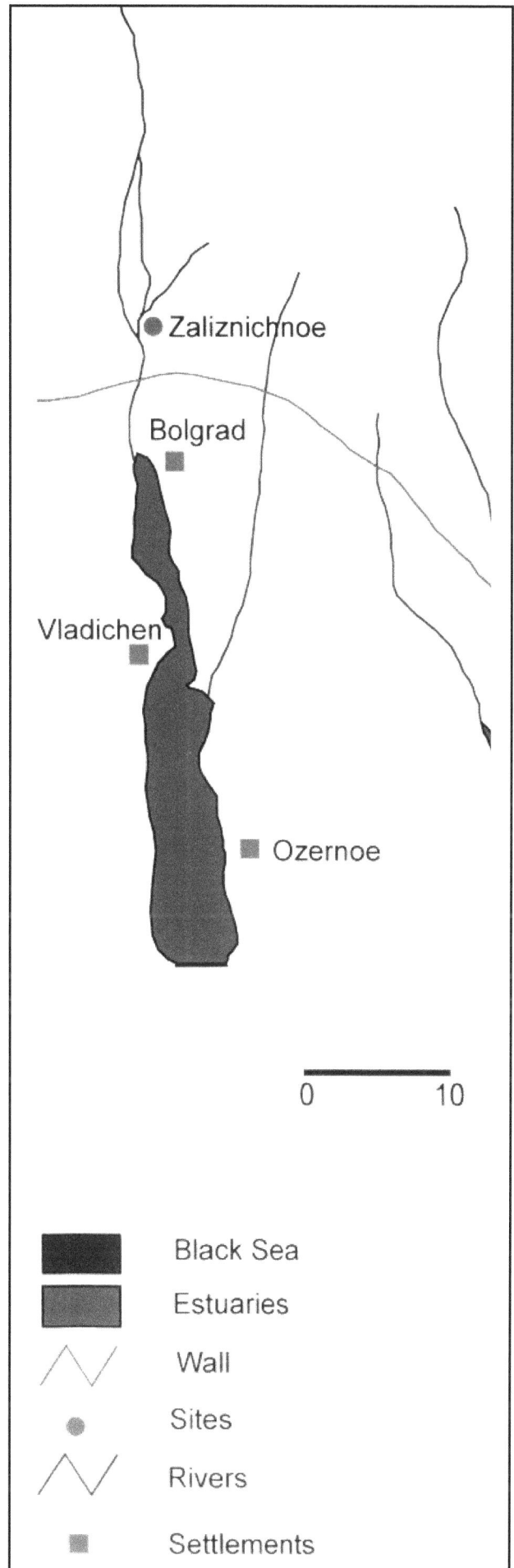

Figure 2. Location of investigated sites.

been studied by Tsalkin (1970) who identified the cattle, sheet/goat, pig and horse. It has been remarked that the cattle at these sites was less common than at early agricultural sites in the South-West Europe. Z.V. Yanushevich (1976) has studied ethnobotanical remains (impressions of grains, fragments of spikelets and glumes on the pottery, mudbricks and plaster) from two Gunmelnitsa sites in Moldavia: Vulcanesti and Lopacica. Among the wheats were identified: emmer (most common), einkorn and spelt; hulled and naked barley, and also oats, millet and blackthorn. The common use of emmer, like that of spelta, was mostly due to their tolerance of poor soil, winter cold and summer draught. The remains of oats were usually found in association with emmer.

BOLGRAD

The site is located on the upper terrace of the Yalpug Liman, on the north-western outskirts of the town of Bolgrad. The settled area lies on the promontory formed by a deep ravine of a small stream, at the altitude of 40 m, close to the abrasion cliff (Fig. 3). The excavated area reached 700 sq. m. The finds of domestic structures were particularly significant (Subbotin 1983). Altogether, the remains of eight dwellings (six of a semi-subterranean type and two surface houses) were reported. The semi-subterranean dwelling No 1 was comparatively well preserved. It formed a large trough-like hollow 7 by 6 m; which floor was found at the depth of 1.6-1.8 below the present-day surface. A circular hearth 1.5 m in diameter was found in the protruding northern part of the dwelling. Another hearth, filled in with potsherds and animal bones, was established in the central part of the dwelling; it formed a cylindrical hole widening at the bottom and reaching a diameter of 0.8 m. The lower part of its filling consisted of cemented ash covered with the layer of mud-bricks, 20 cm thick.

This site included human skeletal remains which belonged to at least three individuals. In one case this was definitely a burial: a well preserved skeleton of a child, 5 - 7 years of age, has been found in a contracted posture together with animal bones and a small polished vessel, the stone slabs both underlying and recovering the body. A large fragment of a skull of an adult individual has been found in the same area, 0.9 m below the previous find. Yet another human skull has been recovered from the same level, 3-4 m to the north-east from the former area.

To this day no more than 30 burials are known for the Gumelnitsa Culture as a whole. In several cases, these are cemeteries consisting of surface graves which were located beyond the dwelling sites. In the majority of cases, the human burials were found either beneath the houses or in the immediate vicinity thereof (Subbotin 1983, 106).

In the course of 1999 field project two sondages, 2 by 2 m. each, were dug, in the northern, previously unexcavated part of the site.

SONDAGE 1

The micellar-carbonatic chernozem loam formed the upper soil level, beneath which, to the depth of 0.65 – 1.00 m, lies the humified sandy loam. At its bottom, sandstone slabs, animal bones and fragments of pottery formed the foundation of the Chernyakhovian semi-subteranean dwelling (Roman Iron Age, 1st-2nd centuries AD), Deeper, at 1.0– 1.35/1.40 m, were found the slabs, animal bones and pottery fragments filling the structure of the sub-semiterranean dwelling of the Gumelnitsa Culture. Two crushed vessels marked the bottom of this dwelling (1.35-1.40 m).

Figure 3. Mesolithic site of Zaliznichnoe.

Figure 4. Gumelnitsa site of Bolgrad; Dr. Subbotin in front of the Sondage 1.

SONDAGE 2

At the depth of 0.70 – 0.85 m several sandstone slabs 5-20 cm in diameter have been found, with numerous Gumelnitsa potsherds and animal bones scattered among them. Within this surface, at the level of 0.80 m, a large fragment of a human skull has been found (Fig. 4). The skull and human bones are presently studied at the Institut de Paléontologie Humaine in Paris.

As preliminary examinations show, this fragment consisted of the parietal bone and the parietal-occipital suture and belonged to an adult individual. The bone tissue was strongly mineralised; with the carbonate crust developed both on the exocranial and endocranial surfaces. Three small holes were visible on the exocranial surface, approximately along a direct line, at irregular distances of 12 and 20 mm. These holes were conic in cross-section, and limited in depth to the outer layer of the bone tissue, not reaching the diploe. They were variably oriented, the smallest, 1.5 mm in diameter, the largest, 3.0 in diameter, was vertical. The holes were apparently human-made, supposedly with the use of a sharp and pointed instrument.

A series of grooves was visible on the bone's surface by side of the smallest hole. Several of these grooves were small and delicate, the others, close to the edge of the broken bone, formed large curvatures, lighter in colour than the rest of the bone tissue. The skull's surface has apparently been scrapped over the length of ca. 3.5 cm, thus forming the traces, about 1 cm wide close to the broken edge, and narrowing to 2 mm at the height of the holes. These traces

cannot be ascribed either to any recognisable anatomical deformity, or to the gnawing by the rodents or any other animals. They can only be due to the human impact with the use of a scraping instrument. One may suggest an intentional removal of the scalp tissue. No similar phenomena had ever been reported from Chalcolithic sites in South-Eastern Europe.

Below this level, at the depth of 0.85 – 1.10 m, one could note numerous potsherds of Gumelnitsa vessels resting on flat sandstone slabs forming a 0.5 m wide pavement, crossing the excavated area in the direction NNE-SSW;

Figure 5. Gumelnitsa site of Bolgrad, Sondage 2, human skull.

At the depth of 1.10 – 1.70 m, two circular hollows filled in with the dark brownish loess were detectable in the western and eastern parts of the sondage; they included the fragments of Gumelnitsa pottery and animal bones. The eastern hollow was 1.00 m in diameter; that of the eastern hollow was 1.60 m in its upper section, narrowing to 1.20 m at the bottom.

As our observations show, the greater part of Gumelnitsa sites in the studied area were located on high levels: the terraced banks of the fresh-water lakes and limans; the banks of the major rivers (the Danube and the Prut), and, in a few cases, the banks of smaller rivers and streams. These naturally fortified settlements had an easy access to arable soils: the carbonate chernozem developed on top of the loess-covered terraces. The largest settlements were in close proximity to the river estuaries ingressed by the sea, which level was at that time by 2-3 m above its present-day position (Fedorov 1978).

The project was sponsored by the INTAS (Grant INTAS-UKRAINE 95-0271).

Authors' addresses

P.M. DOLUKHANOV
University of Newcastle-upon-Tyne
Department of Archaeology
Newcastle, ENGLAND

Michel SÉFÉRIADÈS
CNRS UMR 6566 - Université de Rennes 1
Laboratoire d'Anthropologie, Campus de Beaulieu
F-35042, Rennes, FRANCE

V.N. STANKO
Odessa State University
Department of Archaeology
Odessa, UKRAINE

Bibliography

AKSU A.E., R.N. HISCOTT, D.A. Yasar. 1999. Oscillating Quaternary water level of the Marmara Sea and vigorous outflow into the Aegean Sea from the Marmara Sea - Black Sea drainage corridor. *Marine Geology,* 153, pp. 275-302;

ARTYUSHENKO A.T. 1970. *Rastitel'nost' lesostepi I stepi Ukrainy v chetvertichnom periode.* Kiev, Naukova Dumka.

DUMITRESCU V. 1924. Decouvertes de Gumelnita. *Dacia,* 1, 325-342.

FEDOROV P.V. 1978. *Pleistocen Ponto-Kaspiya.* Moscow, Nauka.

KREMENETSKY K.V. 1991. *Paleoekologiya drevneishih zemledel'cev I skotovodon Russkoi ravniny.* Moscow: Institute of Geography.

PASHKEVICH, G.A. 1982. Paleobotanicheskaya harakteristika poseleniya Mirnoe. In V.A. Stanko *Mirnoe: Problema mezolita stepei Severnogo Prichernomor'ya* (Kiev, Naukova Dumka), 132-8.

ROHLING, E.R. and S. DE RIJK. 1999. Holocene Climate Optimum and Last Glacial Maximum in the Mediterranean: the marine oxygen isotope record. *Marine Geology,* 153, pp. 57-75.

STANKO, V.N. 1982: *Mirnoe: Problema mezolita stepei Severnogo Prichernomor'ya* (Kiev, Naukova Dumka).

SUBBOTIN L.V. 1983. *Pamyatniki kul'tury Gumelnitsa Yugo-Zapada Ukrainy.* Kiev, Naukova Dumka.

TSALKIN V.I. 1970. *Drevneishie domashnie zhivotnye Vostochnoi Evropy.* Moscow, Nauka.

YANUSHEVICH Z.V. 1976. *Kul'turnye rasteniya Yugo-Zapada SSSR po paleobotanicheskim issledovaniyam.* Kishinev, Shtiinca.

POTTERY AND ENVIRONMENT: THE ROMAN PERIOD PRODUCTION CENTRE AT ZOFIPOLE (CRACOW, SOUTHERN POLAND)

H. DOBRZAŃSKA, T. KALICKI, G. CALDERONI & M. LITYŃSKA-ZAJĄC

1. INTRODUCTION

The Przeworsk culture settlement at Zofipole is located within the promontory on the loess terrace of the left bank of the Vistula river, about 28 km east of Cracow. It forms part of a well-defined settlement zone along the Vistula valley. The site is situated 10 m above the flood plain and 1.8 km away from the present river channel. The archaeological materials of the Pre-Roman and Roman Period on the surface of the site extend over 20 ha (Fig. 1). This big site has archaeological material from different phases; at one time period, it thus occupied a smaller area.

The site was surveyed in the course of the 1946-1949 campaign, and then in 1987 and 1996-1997 (Dobrzańska 2000, pp. 37-38). The multicultural settlement at Zofipole was occupied from the Neolithic to the Middle Ages with the predominance of findings from the Pre-Roman and Roman Period which can be dated from the 1st c. BC - 4th c. AD, and particularly to the Late Roman Period (3rd – 4th c. AD), with pottery kilns and remnants of a settlement. Fifty-six kilns located at Zofipole is the largest number ever discovered beyond the frontiers of the Roman Empire in Mid-European Barbaricum.

It is the aim of this paper to present the results of an interdisciplinary research project concerning connections and interactions between man and environment using the example of the Roman Period pottery centre at Zofipole, near Cracow. The task was fulfilled by considering, in addition to archaeological data, information provided by a broad spectrum of disciplines including geomorphology, paleobotany, dendrochronology and paleoclimatology, along with the results of mineralogical and ultrasonic assays and isotope dating.

2. ENVIRONMENTAL SETTING

The Vistula River downstream of the Cracow Gate flows through the western part of the Sandomierz Basin. The erosional relief developed on the Miocene clays was covered by a variety of Quaternary sediments.

On the left bank, there are two Pleistocene Vistula terraces (12 and 15-25 m above the river) covered by Vistulian loess (Fig 1). Within the 8 km wide and 4-5 m above the river flood plain, there are several segments of various ages which record the evolution of the Vistula river valley during the last 13000 years BP (Kalicki 1991). Locally as erosional remnants, relicts of the Young Pleniglacial and Lateglacial braided alluvial plains are preserved. In the remaining area, numerous paleomeanders or their systems of different age occur – large, both Lateglacial and historical ones, as well as several generations of small, Holocene ones. These abandoned channels reflect the lateral migration and avulsions of the Vistula river bed during the Lateglacial and the Holocene. The flood plain is made up of gravel with sands changing upward to sands and silty overbank deposits at the top.

During the Holocene, the Vistula was a meandering, slowly aggradating river (Kalicki 1991). For more than 5000 years, its meandering belt on the study section had a W-E direction and crossed the recent flood plain diagonally (Fig. 1). Very important changes occurred, which consequences place to the time of the Roman period, as well as in the late Atlantic and early Sub-boreal. Most important were the two avulsions of the Vistula river bed at Las Grobla and Zabierzów Bocheński sections, dated to ca. 5000 and 4500 years BP, respectively (Starkel et al. 1991, Kalicki et al. 1996). The flowing directional change of the Vistula River from W-E to SW-NE downstream of Niepołomice was a consequence of the avulsions of river bed to the north (Fig. 1). Besides planar changes, by straightening and shortening the river bed of some tens of kilometres, the avulsions were also responsible for triggering incision (Fig. 2). About 2000-1500 yr BP, the incision reached its maximum, almost 3 m, i.e. the same level as the Alleröd paleomeanders (Kalicki 1991). The meandering and incising river formed the lower level of the flood plain. The top of this was more than one metre below the most recent flood plain (Fig. 3).

This morphological evolution also affected the hydrological conditions of the flood plain by improving its drainage (this causing dropping of site humidity) and increasing the seasonal fluctuations of the ground water level. The climatic instability during the Roman Period also increased flood frequency, bank erosion and the development rate of the point bars. Therefore, a large number of sub-fossil trees dating to this period and buried in alluvia was found (Krąpiec 1998). This very active zone, however, was relatively narrow and limited to the lower level of the flood plain zone. By contrast, on the higher level of flood plain, it is likely that only depressions as paleomeanders were overflooded by water with suspended load and moreover, only during peak discharges (Fig. 3).

Vegetation cover

Research on potential natural vegetation within the area in question has shown that the loess terrace where the settlement was located is habitat to oak and hornbeam forests *Tilio-*

Figure 1. Geomorphological map of the Vistula river valley downstream of Cracow between Branice-Stryjów and Zabierzów Bocheński (by T. Kalicki)
1 – edges below 5 m high, 2 – edges above 5 m high, 3 – palaeomeanders, 4 – small erosional valleys, 5 – Proszowice Upland (Miocene clay hills cover by loess), 6 – Pleistocene Raba alluvial fan, 7 – Pleistocene Vistula's higher terrace (alluvia cover by loess), 8 – Pleistocene Vistula's lower terrace (alluvia cover by loess), 9 – Lateglacial and Holocene Vistula's flood plain, 10 – modern flood plain (inter-dike area), 11 – Lateglacial braiding alluvial plain as backswamps during the Holocene (wide depresion on flood plain), 12 – Holocene alluvial fans, 13 – sloping surface on the convex meander side, 14 – dikes, 15 – limits of meander belt in the Roman time.

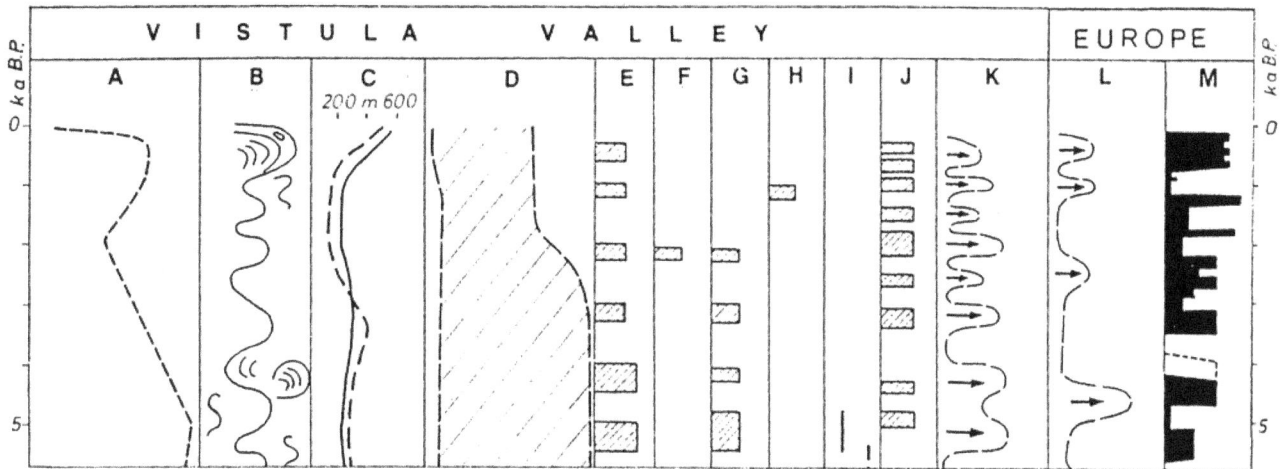

Figure 2. Reflection of the climatic and anthropogenic changes in both morphology and alluvia of the Vistula river valley near Cracow during the last 5000 years (by T. Kalicki)
A-I and K (after Kalicki 1991, modified): A - vertical changes of the Vistula river channel, B - channel pattern changes, cut off and avulsion, C - width (continuous line) and meander radius (broken line) changes, D - mean diameter Mz changes of abandoned channel and overbank deposits, E - cut offs and channel avulsions, F - superposition of the overbank sediments on organic deposits, G - peat initiation, H - cultural horizons beneath the overbank deposits, I - palaeobotanical data indicating a humid (continuous line) and a dry (broken line) climate, J - tree trunks in the alluvia (Krapiec 1996), K - phases of increased activity of the Vistula, L - phases of increased activity of the river in the temperate zone (Starkel 1983), M - lake-level fluctuations in the Jura and sub-Alpine ranges (Magny 1993).

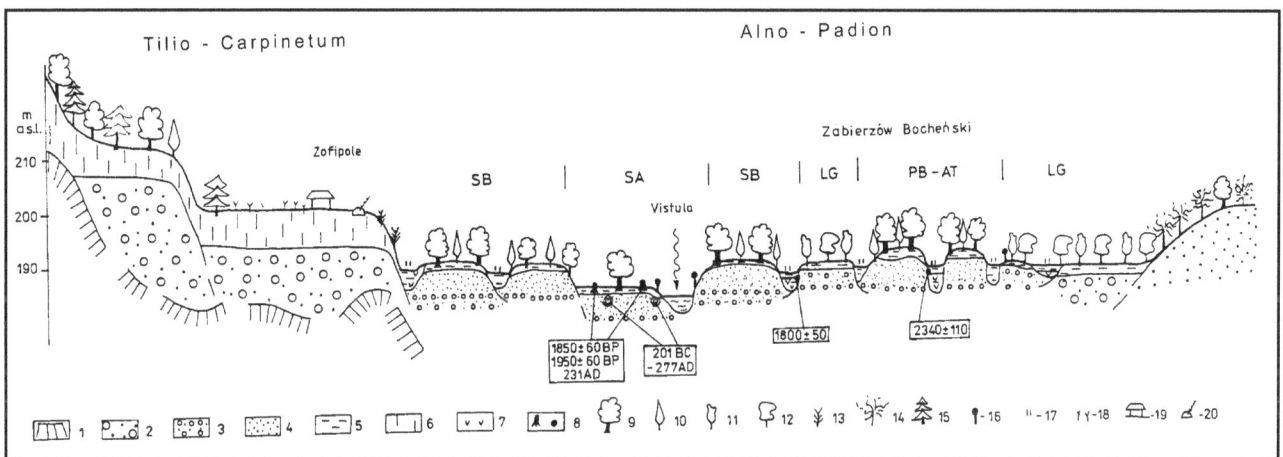

Figure 3. Schematic paleogeographical section across the Vistula river valley in Roman time
(by T. Kalicki with collaboration H. Dobrzańska and M. Lityńska-Zając)
1 – Miocene clay, 2 – Pleistocene gravels with sands, 3 – Holocene gravels with sands, 4 – sands, 5 – silts (overbank deposits), 6 – loess, 7 – peats, 8 – trees and trunks cut by man, 9 – *Quercus* sp., 10 – *Carpinus betulus*, 11 – *Alnus* sp., 12 – *Salix* sp., 13 – *Corylus avellana*, 14 – *Pinus sylvestris*, 15 – *Picea excelsa*, 16 – *Carex* sp., 17 – meadows, 18 – cereals, 19 – settlement: dwelling zone, 20 – settlement: production zone. AT – Atlantic, LG – Lateglacial, PB – Preboreal, SA – Subatlantic, SB – Subboreal. Radiocarbon and dendrochronological datings in boxes.

Carpinetum. These forests, apart from hornbeam *Carpinus betulus* and oak *Quercus robur* and *Q. sessilis*, include lime *Tilia cordata*, and sometimes beech *Fagus sylvatica*, spruce *Picea abies* or silver fir *Abies alba*. Within the lower zone, *Tilio-Carpinetum* forest was in contact with the wetter forest *Alno-Padion* spreading over the Vistula flood plain, and, depending on local conditions, included willow and poplar forests *Salici-Populetum*, alder and ash forests *Circaeo-Alnetum* or elm forests *Fraxino-Ulmetum* as well as groups of reeds, reed swamps and tall sedges. Meadows and wet pastures are the main substitute vegetation groups in this area (Fig. 3).

3. KILN CONSTRUCTION: ENVIRONMENTAL ASPECTS

So far, 35 out of the 56 kilns uncovered at Zofipole centre have been excavated. All of them are completely sunken into the loess and belong to the circular, two-chamber, updraught variety with permanent open-topped firing chamber, in which the lower chamber was divided by a tongue into two sections (Fig. 4). The oven floors were built by plastering the "mada", silty and clay overbank deposits collected from the close flood plain, on a temporary framework of laths, branches and twigs. Among them a variety of species of willow has been

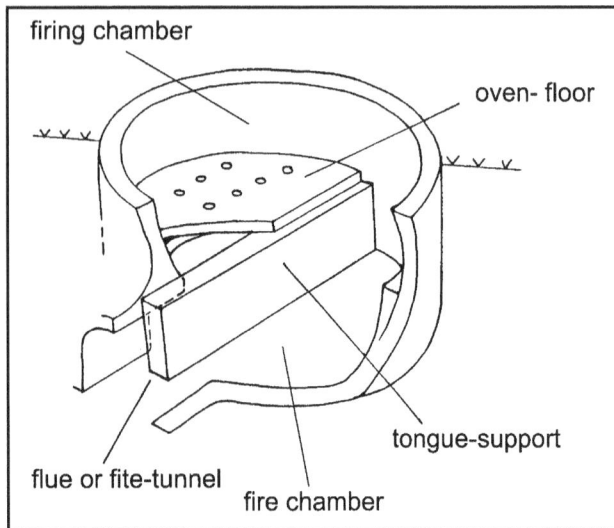

Figure 4. Pottery kiln with permanent open-topped superstructure.

identified, used to grow on the flood plain. Also, the imprints of burdock leaves prove that material was obtained from the lower scrub zone near the Vistula.

The spatial distribution of the kilns shows that their location matches the most favourable geomorphologic context. All kilns were cut on the edge of the upper loess terrace of Vistula river valley (Fig. 5) and, in most cases, the stokehole bottom is deeper than the associated kiln. As all the kilns are completely sunken into the loess, no internal plastering was needed and clay was used only to build the oven floors.

Sunken kilns ensured a perfect oxygen-free reducing atmosphere of vessel firing. Since kilns were built in the loess and were well isolated, they were the preferred type of construction. An analysis of the arrangement of the kilns has shown that the inlets of the fire tunnels were never situated to the SW and W, due to the prevailing wind directions in this region (Fig. 6). A sudden gust of wind, carrying oxygen, was a danger to reducing firing. However, rain and fog did not adversely affect the firing process, which is why the kilns were never sheltered.

4. VESSELS

The potters of Zofipole produced two basic types of grey pottery: fine "tableware" and coarse "kitchenware". Laboratory analysis of clay samples compared with pottery testing has revealed that the principal raw material available was "mada" taken from the flood plain of the Vistula River. Both combined coiling and wheel-thrown techniques were used to make fine and coarse pottery (Dobrzańska, Piekarczyk 1999-2000). Ceramics were fired in a reducing atmosphere, called hydrothermal firing (under the pressure of water vapour). With this method, the average firing temperatures were approximately 700°C ±35, while the maximum temperature never exceeded 800°C (Dušek, Hohmann, Müller, Schmidt 1986). The kilns of Zofipole were heated

with oak wood (*Quercus sp.*) whereas alder (*Alnus*), pine (*Pinus sylvestris*), birch (*Betula*), silver fir (*Abies alba*) and willow (*Salix*) were quite rare among the remains from the kilns. One of the advantages of oak is that it is a highly caloric wood, and therefore burns slowly with a low flame and produces smoke. These properties are essential for grey pottery firing. In order to burn leaves that contained water, oak, alder and birch branches and twigs were used in the hydrothermal process.

5. SPATIAL STRUCTURE AND ORGANISATION OF THE SITE

All the kilns found at Zofipole were built along the edge of the high loess Vistula terrace. They are concentrated in the western section of the site. Both inner chronology and lifespan of the pottery centre rely on critical evaluation of the results of archaeological analyses and radiocarbon dating of well preserved charcoal fragments strictly associated with the kiln activity. They prove that kilns were not operating simultaneously between 200-375 AD (Dobrzańska 2001, pp. 47-49). The remnants of this settlement have been discovered north of the edge of the Vistula loess terrace (Fig. 5).

Due to the fact that the excavations conduced on site in the past aimed almost solely at identifying the "kiln area", it is virtually impossible to reconstruct this settlement in greater detail. One of the activities of the local population, indicated by plant impressions in the clay grates of kilns, was farming. In the Roman Period, barley *Hordeum vulgare* and millet *Panicum miliaceum* were the most popular cereals. Emmer wheat *Triticum dicoccon* has been found in a number of archaeological sites of the Przeworsk culture, although its role was slightly less significant than in older periods. Imprints of grains and pieces of cereal straws and pea seeds suggest that these plants were grown locally. Iron pieces of a wooden shovel plough also indicate farming. Numerous bones of domestic animals clearly indicate that the inhabitants bred livestock, and breeding might have played a significant role in their economic activities (Dobrzańska 2000, p. 62).

In the production area, beside the pottery manufacturing, remnants of bronze-smiths' workshops have also been discovered with a chronology corresponding to the pottery production. The production zone discussed here was a "fire activity" zone, and was separated from the rest of the settlement for safety reasons.

6. DISCUSSION

The pottery site at Zofipole existed during the period 2350-1800 BP within a generally cooler and more humid climate (Kalicki 1996b, Frenzel 2000). Traces of this period can be found in different enviroments not only in Poland but in all of Central Europe and also in Northern Europe.

The high level of Polish and Subalpine lakes occurred (Ralska-Jasiewiczowa 1989, Magny 1993). The mountain glacier advance phases in the Alps and in the Scandinavian

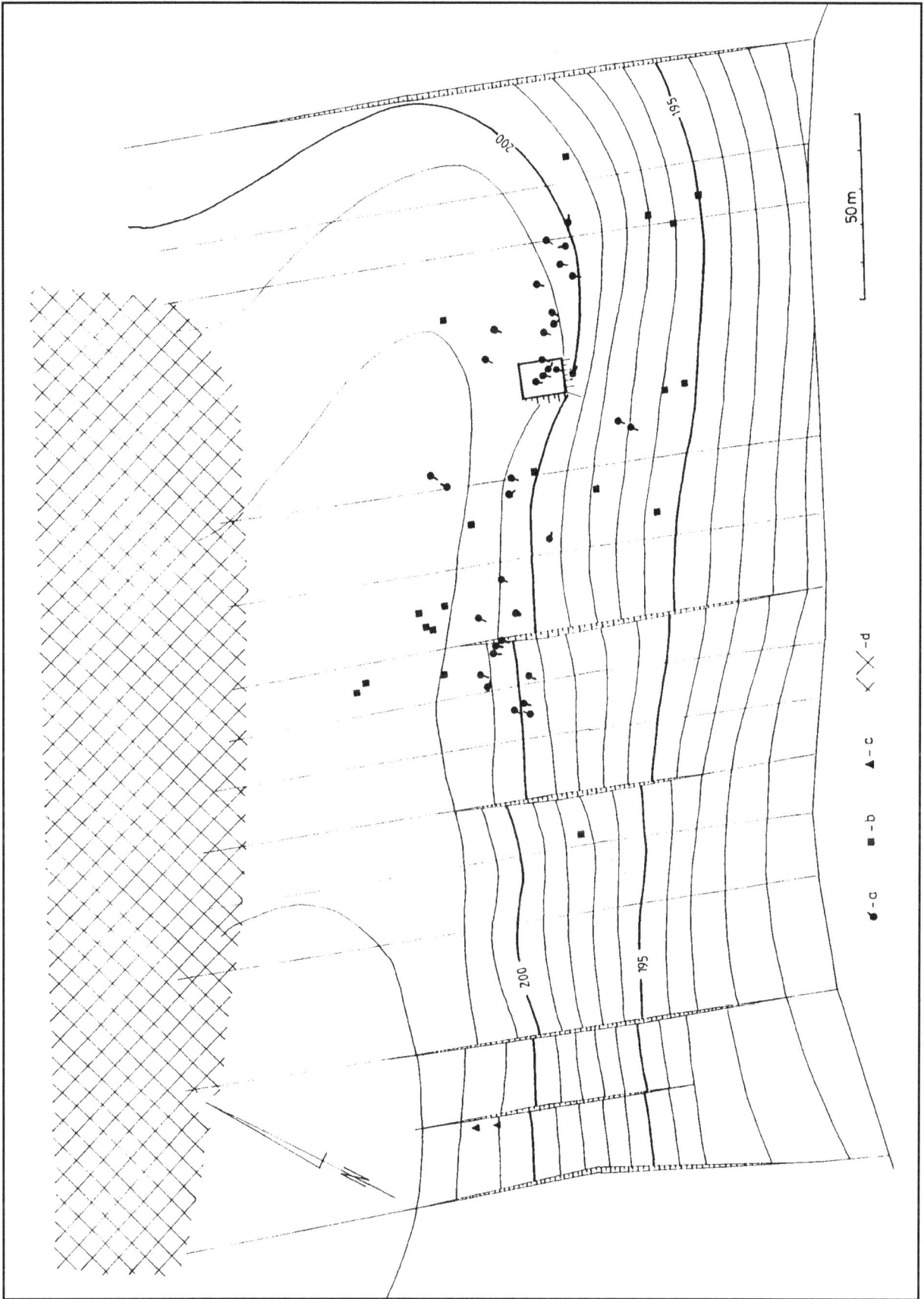

Figure 5. Zofipole, site 1: a – excavated kilns, b – position of kilns based on the magnetic prospection, c – kilns uncovered as result of the archaeological survey, d – housing area.

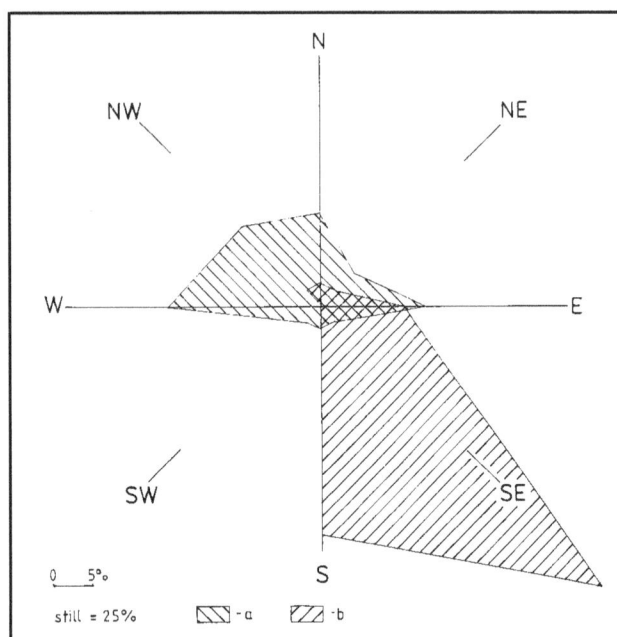

Figure 6. Relationship between distrribution of the wind directions (Hess 1968, p. 74, tab. 26) (a) and position of the fire tunnel inlets (b).

Mountains, development of nival processes in the Ukraine Carpathians, solifluction in the Swiss Alps and landslides in the Pyrenees, the Cantabrian Mountains and in Great Britain took place at this time (see references in Margielewski 1998). An increase in river activity in numerous valleys (Rhine and its tributary in Danube valley and its Alpine tributaries, both in the valley and drainage basin of Rhone) also occurred (see references in Kalicki 1996b and Dobrzańska, Kalicki in press). The cooling and more humid climatic phase in more northern and oceanic regions of Europe also occurred. It led to the development of peat bogs in numerous paleochannels of the Pulmankijoki in Finland and to alluviation and braiding of channel beds in Great Britain (see references in Dobrzańska, Kalicki in press). Similar climate changes are observed in northern Poland. It led to the development of intensive fluvial processes in the Pomeranian valley, an increase of the ground water level and development of large peat bogs in the Gardno-Łeba Lowland and renewed water pools in the depression on the Słupia's overflood terrace and renewed water in Gardno Lake (Florek 1989). Also in central Poland in the Moszczenica valley, a settlement of Przeworsk culture (250-350 AD) was covered by overbank deposits, and about 1800±80 BP (cal. 130-340 AD) began aggradation which led to fossilisation of soil covered by channel deposits (Kamiński, Moszczyński 1996). Regional differences probably were present during the Roman Period. The reduction of the lake level in central Poland indicates a dryer and warmer climate between 2050 and 1750 BP (Wojciechowski 2000). This climatic oscillation occurred during a period with significant anthropogenic changes of the Central European natural environment; traces of this climatic phase are thus more distinct than in the early and middle Holocene (Kalicki 1996a).

Abundant data from the different geo-systems also reflect the climatic destabilisation during this epoch: an increase in the frequency of extreme events due to various factors. The data from southern Poland indicate that in the studied period, events associated with both marine and continental climate changes took place. Often rainy periods, some days with rain after advection of humid air masses from the west, result from marine climate change and contributed to floods in the upper Vistula drainage basin and landslides in the Flysh Carpathians. (Kalicki 1991, Margielewski 1998) Sub-fossil trees in alluvia (Krąpiec 1998) document bank erosion during these floods. The short rainfalls caused flash floods in small valleys (Alexandrowicz 1997) and debris flows on high mountain slopes (Baumgart-Kotarba, Kotarba 1993) are typical of continental climate changes.

Despite these often unprofitable climatic events during this period, people managed to intensively occupy not only overflood terraces, but also penetrated the flood plain. This could be related to local conditions, the incision of the Vistula channel and the limiting of high river activity to only the relatively narrow zone of lower flood plain and to good drainage of the remaining valley floor (Fig. 3). Human penetration of the flood plain was periodical due to frequent inundation which made more stable use impossible. Inhabitants of the Zofipole settlement obtained wood for fuel and construction from the valley floor. Traces of this are trunks of oaks cut by man dating to 2260±80 BP, 1950-1850 BP cal. 35 BC-240AD (Kalicki, Krąpiec 1991). They were found on the remnant of the fossil, lower flood plain at Branice-Stryjów (Kalicki, Krąpiec 1995). At that time, people probably mainly deforested the lower flood plain and also used trees fallen naturally by bank erosion of the river because the river channel was the most convenient "road" on the flood plain. The trees could be floated to the vicinity of Zofipole because the meander belt from Roman times approached the edge of the loess terrace (Fig. 1).

These favourable local environmental conditions enabled the sustained development of settlement in Zofipole. Beneficial for growing crops were fertile soils (chernozems) on loess terraces, while the pottery production was based on easily accessible good quality loams and woods, overbank deposits and oak forest on the flood plain. The oak woods provided material for construction and fuel in other areas of manufacture (bronze and iron smithy). Wood was also extensively used in the manufacture of tools and other items of daily life and as a means of transport, e.g. a canoe hollowed out of an oak tree trunk found in the Vistula alluvia a few kilometres east of Zofipole, dendrochronologically dated to 212 AD (Krąpiec 1996). The proximity of the river also had an impact on the development of the settlement. The river not only supplied water to the inhabitants but also enabled a number of activities related to the already mentioned manufacture. The discovery of the wooden canoe suggests that the river was used for transport, probably also for transportation of goods, and, perhaps, primarily of those made in the settlement. It should be emphasised that water transport was the cheapest means of transport.

Zofipole was a village which pursued a multidirectional economy. The present state of examination of the settlement does not allow us, however, to determine the proportion of its particular elements. Given the results of analysis of the natural environment and discovery of archaeological artefacts, and paleobotanical and paleozoological remains, we may assume that at Zofipole, inhabitants lived primarily by farming. Similarly to the neighbouring Igołomia (Dobrzańska 1990, pp. 91-100), the volume of workshop production exceeded that of the needs of the inhabitants of a single settlement and should be regarded as the main subsistence strategy of the potters.

Grey pottery was made at Zofipole settlement for about 180 years. This production centre ceased to exist in the last quarter of the 4th century, due to the collapse of the Przeworsk culture within the area east of Kraków, which is confirmed by relevant archaeological sources. The end of human economic activity in this area led to the regeneration of oak on the Vistula flood plain in the late 4th century and at the beginning of the 5th c. AD (Fig. 7). This type phase of oak germination starting at 400 AD was also distinguished by Leuschner *et al.* (2000) in the Main and Danube valleys, particularly in connection with human impact changing the rivers' regime. Data from the Vistula valley could indicate rather an association with direct discontinuous or distinctly less clearing human activity on the flood plain.

This economic decline was connected with changes in the political situation in European Barbaricum, especially with the invasion by the Huns and the beginning of Goth migrations (ca. 375 AD). These disturbances and wars caused destabilisation of the economic situation. The "settlement hiatus" in the Cracow region should be interpreted as a significant decline in settlement density and fall of some of the so-called craft branches (Dobrzanska 2000, pp. 62-63).

7. CONCLUSIONS

1. The village at Zofipole forms part of a well-defined settlement zone about 30 km long, along the Vistula valley, east of Cracow. In that area eight larger villages of the Pre-Roman and Roman Period were situated, of the same economic structure (production zones with pottery workshops and bronze and iron smithy) and chronology comparable to Zofipole (Dobrzańska 1990, pp. 104-113, Fig. 27). This kind of economy created a great demand for wood resources.

2. The site at Zofipole may date from the Pre-Roman to the Roman Period (2nd c. BC to the third quarter of the 4th c. AD). An intensive settlement and economic activity of its inhabitants can be observed from the second half of the 2nd c. AD to the last quarter of the 4th c. AD (especially between 200-375 AD) and may be correlated with the lack of oak growth on the flood plain (Fig. 7).

3. People used the flood plain nearly all year long because floods are passing events and local conditions such as the incised river causing a dry, good draining valley bottom were favourable for human occupation. The many trees felled naturally by bank erosion in huge floods were also easily accessible raw material.

4. The decline of settlement in the Vistula valley east of Cracow was favourable for the reforestation of oaks on the flood plain (Fig. 7). Synchronous declining phases of oaks in Central European river valleys thus reflect climate changes (see Kalicki, Krąpiec 1995), but the late Holocene phases of oak germination on flood plains reflect human impact. Reforestation was connected not with river regime changes as suggested Leuschner *et al.* (2000), but with a decline in human felling activity.

5. In the studied area, an increase of human activity coincided with a period of numerous floods and other catastrophic events in relatively cooler and humid phases. On the contrary, in the period of the warmer climate, human activity declined (i.e. "settlement hiatus" from AD 375 to the second half of the 5th c. AD).

Authors' addresses

H. DOBRZAŃSKA
M. LITYŃSKA-ZAJĄC
Institute of Archaeology and Ethnography
Polish Academy of Sciences
ul. Slawkowska, 17, 31-316, Kraków, POLAND

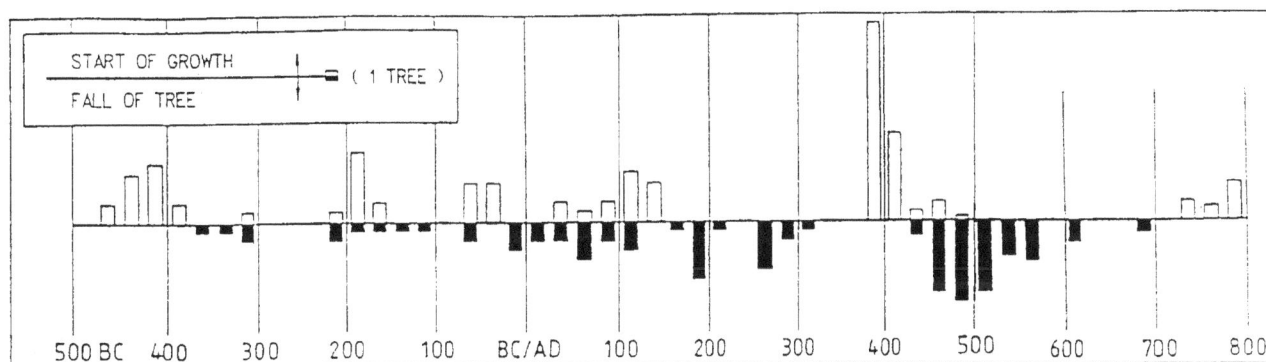

Figure 7. Phases of beginning of the growth and felling of the oaks from the Vistula river valley near Cracow from 500 BC to 800 AD (by M. Krąpiec in Kalicki, Krąpiec 1996).

T. KALICKI
Instytut Geografii i Przestrzennego Zagospodarowania PAN
ul. Św. Jana 22 31-018, Kraków POLAND

G. CALDERONI

Bibliography

ALEXANDROWICZ S. W., 1997, Malacofauna of Holocene sediments of the Prądnik and Rudawa river valleys (southern Poland), "Folia Quaternaria" 68, 133-188.

BAUMGART-KOTARBA M., KOTARBA A., 1993, Późnoglacjalne i holoceńskie osady z Czarnego Stawu Gąsienicowego w Tatrach (in:) A. Kotarba ed., Z badań fizyczno-geograficznych w Tatrach "Dokumentacja Geograficzna" 4-5, 9-30.

DOBRZAŃSKA H., 1990, Osada z późnego okresu rzymskiego w Igołomi, woj. krakowskie. Część II (The Late Roman Period Settlement at Igołomia, Cracow Province. Part II) Kraków.

DOBRZAŃSKA H., 2000, Ośrodek produkcji ceramiki "siwej" z okresu rzymskiego w Zofipolu (Das Produktionszentrum der kaierzeitlichen "grauen" Keramik in Zofipole) (in:) J. Rydzewski ed., 150 lat Muzeum Archeologicznego w Krakowie, Kraków, 37-68.

DOBRZAŃSKA H., Kalicki T., in print, Interakcja człowiek-środowisko w dolinie Wisły koło Krakowa w okresie od I do VII w n.e. (in:) Koszęcin.

DOBRZAŃSKA H, PIEKARCZYK J, 1999-2000, Ultrasonic testing of ceramic vessels of Roman Period from the production centre at Zofipole, "Acta Archaeologica Carpathica" 35, 89-111.

DUŠEK S., HOHMANN H., MÜLLER W., SCHMIDT W., 1986, Rekonstruktion eines Töpferofens und des Brennvefahrens, Weimarer Monographien zur Ur- und Frühgeschichte 16, Weimar.

FLOREK W., 1989, Rozwój sieci rzecznej Przymorza na przełomie okresów subborealnego i subatlantyckiego a działalność człowieka (na przykładzie Słupi i Łupawy), (in:) Problemy kultury łużyckiej na Pomorzu, 173-184, Słupsk.

HESS M.,1968, Klimat terytorium miasta Krakowa, "Folia Geographica, Series Geographica- Physica" 1, 35-97.

KALICKI T., 1991, The evolution of the Vistula river valley between Cracow and Niepołomice in late Vistulian and Holocene times, (in:) Evolution of the Vistula river valley during the last 15 000 years, part IV, "Geographical Studies", Special Issue No. 6, 11-37.

KALICKI T., 1996a, Climatic or anthropogenic alluviation in Central European valleys during the Holocene (in:) J. Branson, A. G. Brown, K. J. Gregory red., Global Continental Changes: the Context of Palaeohydrology, Geological Society Special Publication 115, The Geological Society, 205-215, London.

KALICKI T., 1996b, Phases of increased river activity during the last 3500 years, (in:) Evolution of the Vistula river valley during the last 15 000 years, part VI, "Geographical. Studies", Special Issue No 9, 94-101.

KALICKI T., KRĄPIEC M., 1991, Black oaks and Subatlantic alluvia of the Vistula in the Branice-Stryjów near Cracow, (in:) Evolution of the Vistula river valley during the last 15 000 years, part IV, "Geographical Studies", Special Issue No. 6, 39-61.

KALICKI T., KRĄPIEC M., 1995, Problems of dating alluvium using buried subfossil tree trunks: lessons from the "black oaks" of the Vistula Valley, Central Europe, "The Holocene" 5, 2, 243-250.

KALICKI T., KRĄPIEC M., 1996, Reconstruction of phases of the "black oaks" accumulation and of flood phases, (in:) Evolution of the Vistula river valley during the last 15 000 years, part VI, "Geographical Studies", Special Issue No. 9, 78-85.

KALICKI T., STARKEL L., SALA J., SOJA R., ZERNICKAYA V. P., 1996, Subboreal paleochannel system in the Vistula valley near Zabierzów Bocheński (Sandomierz Basin), (in:) Evolution of the Vistula river valley during the last 15 000 years, part VI, "Geographical. Studies", Special Issue, No. 9, 129-158.

KAMIŃSKI J., MOSZCZYŃSKI J., 1996, Wpływ osadnictwa kultury przeworskiej na kształtowanie doliny Moszczenicy w okolicy Woli Branickiej, "Acta Geographica Lodziensia" 71, 55-66.

KRĄPIEC M., 1996, Dendrochronology of "black oaks" from river valleys in Southern Poland, (in:) Evolution of the Vistula river valley during the last 15 000 years, part VI, "Geographical. Studies", Special Issue No 9, 61-78.

KRĄPIEC M. 1998, Oak dendrochronology of the Neoholocen in Poland, "Folia Quaternaria" 69, 5-133.

LEUSCHNER H. H., SPURK M., BAILLIE M., JANSMA E., 2000, Stand dynamics of prehistoric oak forest derived from dendrochronologically dated subfossil trunks from bogs and riverine sediments in Europe, "Geolines" 11, 118-121.

MAGNY M., 1993, Holocene fluctuations of lake levels in the French Jura and sub-Alpine ranges and their implications for past general circulation patterns, "The Holocene" 3, 4, 306-313.

MARGIELEWSKI W., 1998, Landslide phases in the Polish Outer Carpathians and their relation to climatic changes in the Late Glacial and the Holocene, "Quaternary Studies in Poland" 15, 37-53.

RALSKA-JASIEWICZOWA M. ed., 1989, Environmental changes recorded in lakes and mires in Poland during the last 13 000 years, III, "Acta Palaeobotanica" 29.

STARKEL L., 1983, The reflection of hydrologic changes in the fluvial environment of the temperate zone during the last 15 000 years (in:) K. J. Gregory red. Background to palaeohydrology: a perspective, J. Wiley, 213-237, Chichester.

STARKEL L., GĘBICA P., NIEDZIAŁKOWSKA E., PODGÓRSKA-TKACZ A., 1991, Evolution of both the Vistula floodplain and lateglacial-early Holocene palaeochannel systems in the Grobla Forest (Sandomierz Basin) (in:) Evolution of the Vistula river valley during the last 15 000 years, part IV, "Geographical Studies", Special Issue No. 6, 87-99.

WOJCIECHOWSKI A., 2000, Zmiany paleohydrologiczne w środkowej Wielkopolsce w ciągu ostatnich 12 000 lat w świetle badań osadów jeziornych rynny kórnicko-zaniemyskiej, Seria Geografia 63, Wydawnictwo Naukowe UAM, Poznań.

NEW PALEONUTRITIONAL DATA FROM MEDITERRANEAN PREHISTORIC POPULATIONS: AN ATTEMPT AT INTERPRETATION OF THE CHANGES IN DIET AND ECONOMY THROUGH TIME

Fulvio BARTOLI, Barbara LIPPI, Emiliano CARNIERI,
Francesco MALLEGNI & Francesca BERTOLDI

Summary: We present here a paleonutritional picture of ancient Mediterranean prehistoric populations from Upper Paleolithic to Iron Age. The samples have been analysed through AAS in order to distinguish between a protein-based diet and a high consumption of cereals or fish and therefore between H/G's economies and exploitation of agricultural resources of more advanced societies.

Resumè: Dans ce travail nous avons essayé de donner une interprétation des dates recueilli dans nombreux années des études de paleonutrition, par les analyses du AAS. Les analyses sont été fait au bout de 1990, dans le laboratoire des Etudes de Paleonutrition de l'Université de Pisa. Les éléments en trace analysé sont Sr et Zn. Ils sont indicateurs d'une haute incidence dans l'alimentation de végétales, céréales et poisson (dans le cas du Sr) ou de lait et viande rouge (dans le cas du Zn). Dans les plus récentes échantillons nous avons essayé d'analyser aussi Cu, Mg, Ba, V par compléter le notre tableau et pour distinguer entre la subsistance relative à l'agriculture et à l'élevage des bestiaux. Ces données peuvent être mis en relation avec ceux qui regardent les restes faunistiques et botaniques et ceux qui regardent les manufactures dans le but d'une meilleure compréhension de l'utilisation des ressources nutritionnelles pendant la transition du Paléolithique au Mésolithique et au Néolithique (des sociétés H/G à celles agricoles). Un des rôles plus important dans les communautés Italiennes a été joué par l'élevage et la pêche. Notre analyses nous avons porté à prendre en considération aussi la période du Néolithique jusque à l'age du fer, et nous montrent la naissance des premières sociétés complexes et stratifiés et le debout du procès d'urbanisation.

Since the UISPP Congress held in Forli in 1996 we decided to enlarge our sampling of prehistoric populations in order to build up a picture of the paleonutrition of ancient Italian populations as detailed as possible and to fill the gaps regarding the most ancient period in prehistory: the Paleolithic.

We divided and presented our samples following the chronological distinctions among Paleolithic (Gravettian and Epigravettian samples), Natufian and Neolithic, Copper and Bronze age (Early, Middle, Late and Final) and Iron age, embracing in this period also samples belonging to the first Italic historical cultures.

The samples have been analysed through AAS in the Paleonutrition laboratory of Pisa University and the trace elements that have been taken into consideration are Strontium and Zinc, as indicators of a diet rich in vegetables, cereals or fish or a diet rich in animal proteins such as red meat and milk. Obviously samples presenting a diet rich in Sr or Zn will allow us also to dicriminate-together with the evidence offered by other kinds of archaeological evidence-among the ways of subsistence and economies on which the human group was relying: large animals-hunting, H/G way of living, plant and animal husbandry, exploitation of fish resources...

The results obtained for Paleolithic samples have been presented as single data given the scarcity of human remains dating back to this period, while for other periods we considered mean values.

Paleolithic individuals come from Central and Southern Italy and have been divided into Gravettian and Epigravettian samples (graph 1 and table 1). The most ancient ones are represented by the samples from Paglicci and Parabita while the most recent ones include the group from Grotta Continenza (nine individuals), Romanelli, Grotta del Romito and Vado all'Arancio.

We can immediately notice the high values of Zn/Ca ratio in the most ancient ones, a value that becomes definitely lower in most recent samples that show thus an inverted ratio between Zn/Ca and Sr/Ca. The evidence offered by our tests seems to confirm a high consumption of animal proteins derived from hunting activities in the most ancient phases of Paleolithic that decreases in later times with a diet that appears as less adequate and richer in foods rich of Sr, such as fish, molluscs, and wild vegetables.

In Neolithic times Sr values are high while Zn ones are generally below standard ones, thus indicating the introduction of cereals and vegetables and at least in coastal or lakeshore sites the importance of fish resources in Neolithic times as it has been confirmed by faunal analysis of several sites.

Comparison data have been obtained from the literature and regard Sr values of Natufian sites that appear as homogeneous with our samples from Central and Southern Italy (graph 2 and table 2).

Values of Sr/Ca and Zn/Ca in samples dating to Copper and Bronze age are presented in graph 3 and table 3.

Table 1. Values of Sr/Ca and Zn/Ca in Paleolithic samples.

	Sr/Ca	**Zn/Ca**
Paglicci 12	1,04	1,72
Paglicci 25	0,93	1,73
Parabita 1	0,5	0,93
Parabita 2	0,7	1,5
Vado all'Arancio A		1,1
Romanelli	0,76	0,95
Romito 1	0,53	1,08
Romito 2	0,37	0,69
Romito 3	1,48	1,3
Romito 4	0,66	0,4
Romito 5	0,8	0,83
Romito 6	0,74	0,67
Continenza 1	0,46	0,33
Continenza 2	0,43	0,2
Continenza 3	0,75	0,23
Continenza 4	0,76	0,27
Continenza 5	0,74	0,44
Continenza 6	0,76	0,34
Continenza 7	0,72	0,2
Continenza 8	0,4	0,4
Continenza 9	0,61	1,09

Table 2. Average values of Sr/Ca and Zn/Ca in Neolithic samples.

	Sr/Ca	**Zn/Ca**	No. of individuals
Haji Firuz	0,68±0,23		16
Ganj Dareh	0,65±0,27		16
Hayonim	0,78±0,10		14
El Wad	0,92±0,20		21
Kebara	0,75±0,17		14
Mallaha	0,56±0,16		10
Nahal Oren	0,61±0,22		20
Nahal Oren	0,91±0,19		10
Catignano	0,75	0,29	1
Continenza M	0,66	0,36	2
Continenza F	0,82±0,03	0,45±0,06	3
Continenza ?	0,86	0,43	2
Samari M	0,82±0,14	0,47±0,10	7
Samari F	0,56	0,62	2
Ripa Tetta M	0,74	0,35	2
Ripa Tetta	0,63	0,32	1
Latronico	0,56	0,31	1
Trasano M	0,79	0,45	2
Trasano F	0,82	0,47	2
Pulo di Molfetta	0,92	0,31	1
Balsignano	0,81	0,52	1
Monte Kronio	0,76	0,46	1

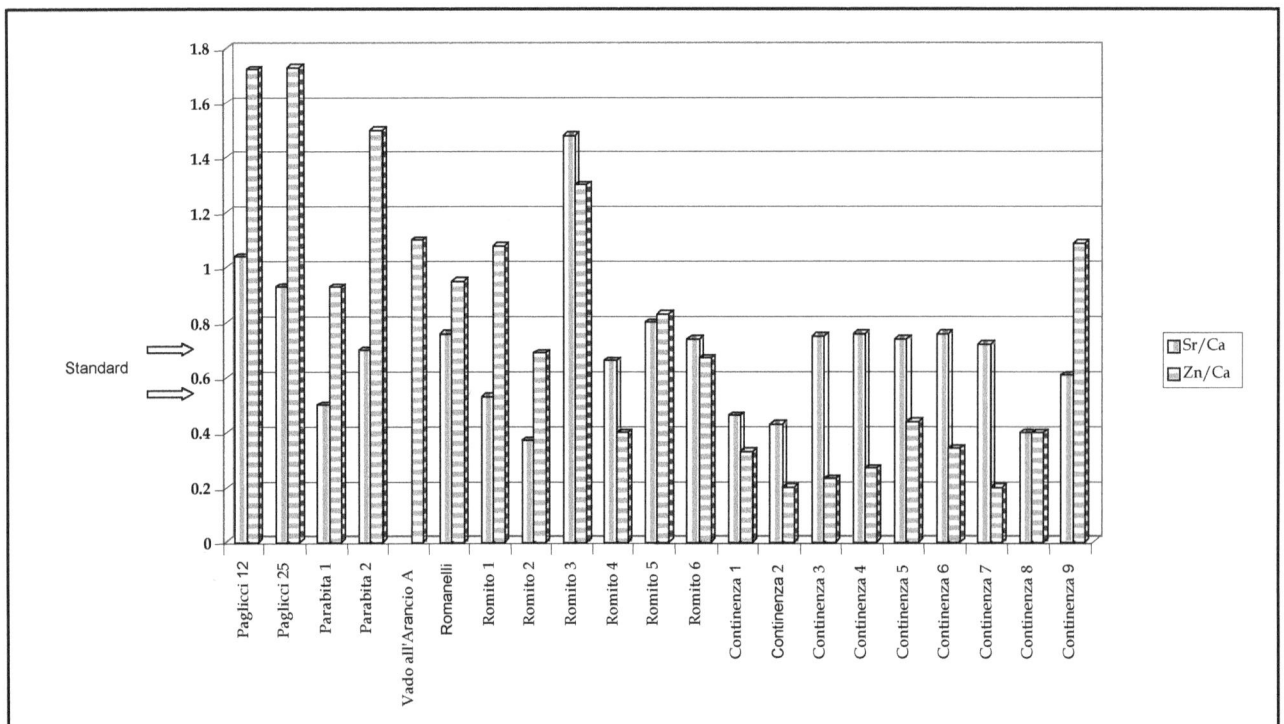

Graph 1. Upper Paleolithic samples (Sr/Ca; Zn/Ca)

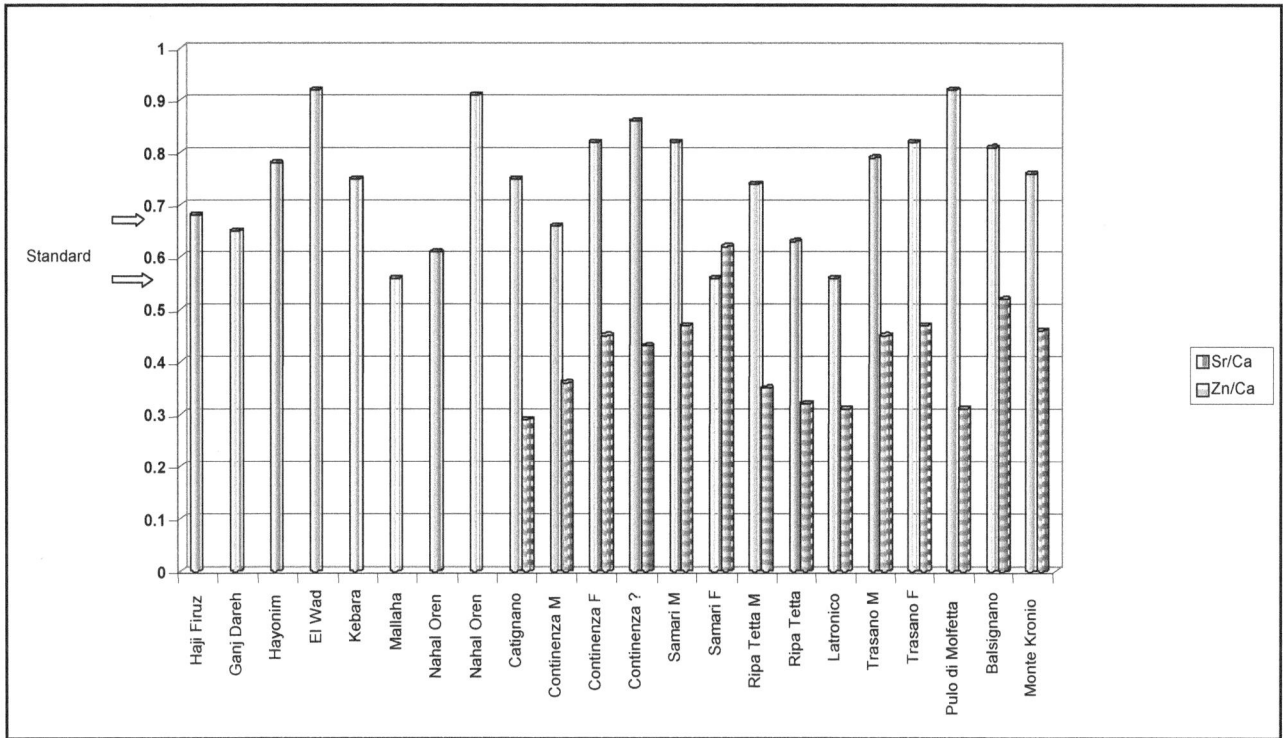

Graph 2. Neolithic samples (Sr/Ca; Zn/Ca)

Table 3. Average values of Sr/Ca and Zn/Ca in Copper and Bronze age samples.

	Sr/Ca	Zn/Ca	Number of individuals
Grotta San Giuseppe (LI)-Copper age	0,42±0,09	0,56±0,13	46
Grotta Prato (GR)-Copper age		0,51±0,12	29
Piano di Sorrento (NA)-Copper age	0,75±0,11	0,51±0,08	13
Piano Vento (AG)-Copper age	0,79±0,19	0,66±0,30	32
Abisso Cesca (TS) Bronze age	0,13	0,83	2
Morano Po (AL)-Late Bronze age	0,6±0,17	0,11±0,06	20
Monteceti (AN)-Bronze age	0,67±0,19	0,18±0,06	3
Monte Fulcino (LT)-Bronze age	0,31±0,03	0,67±0,20	34-29
Celano (AQ)-Late Bronze age		0,78±0,20	3
Trinitapoli sett. AB (FG)- Bronze age	0,55±0,1	0,43±0,14	32
Trinitapoli sett. C (FG)-Bronze age	0,71±0,16	1,08±0,27	32
Grotta Llardo (CT)-Bronze age	0,42±0,11	0,47±0,07	3
Casa dello Studente (ME)-Bronze age		0,61±0,40	6
Ribera Ciavolaro (AG)-Bronze age	0,53±0,12	0,57±0,14	30
Kalinkaya 1-Turkey	0,71±0,22	0,61±0,19	10
Pico Ramos-Spain	0,74±0,53	0,38±	25
Kalinkaya 2-Turkey	0,80±0,36	0,72±0,59	32
Karatash-Turkey		0,44±0,16	10
Manika-Greece	0,51±0,16	0,51±0,20	25
Nichoria-Greece	0,55±0,10	0,38	26

Values of Zn/Ca and Sr/Ca generally appear as adequate to nutritional standards and Zn values in Italian specimens are generally higher than in other Mediterranean countries. The economy was probably mixed and relying on agriculture and animal husbandry as testified by archaeological evidence (introduction of the plough since Copper age, intense animal

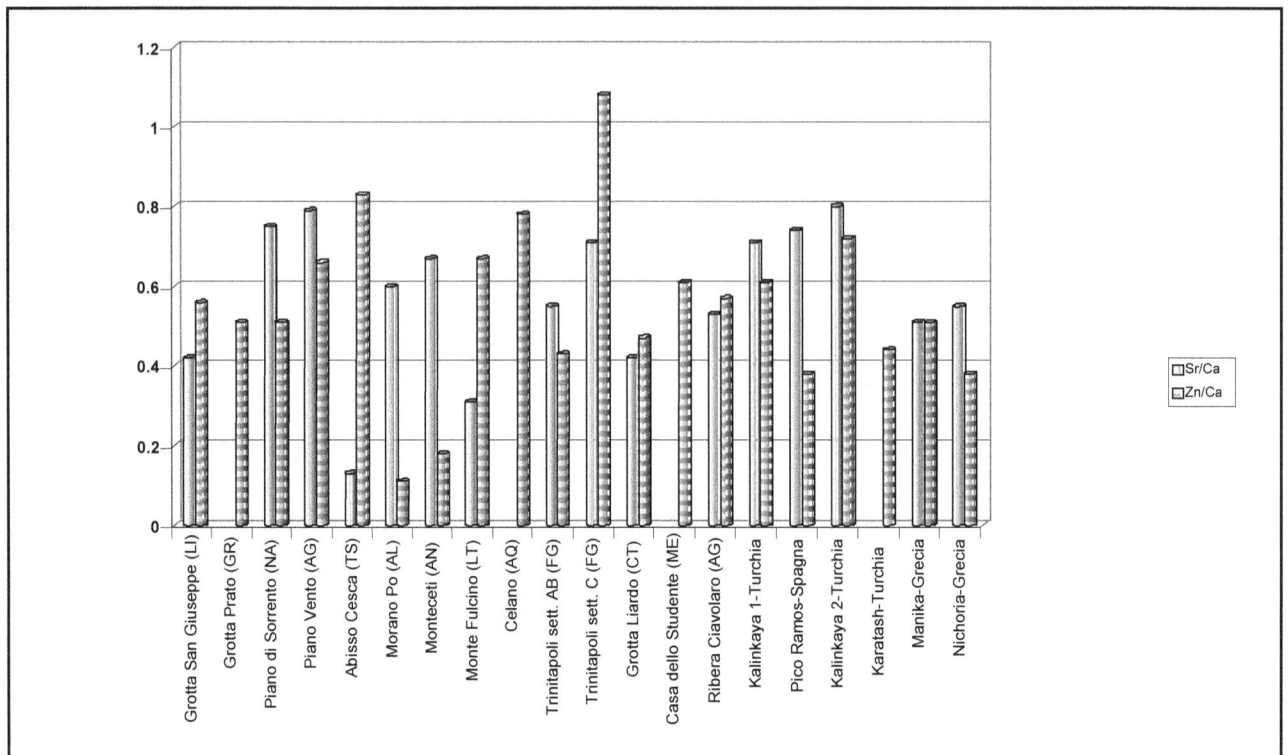

Graph 3. Copper and Bronze age samples (Sr/Ca and Zn/Ca)

husbandry, clearance of wood areas to get larger pastures, highland zone exploitation, appearance of peculiar pottery shapes...)

Iron age samples show (graph 4 and table 4) a surprising lower Zn/Ca ratio that in most of the cases falls well below the standard values thus maybe indicating the effects of the crisis that took place during Late and Final Bronze age: agriculture seems to become the most important economic resource.

Acknowledgements

We wish to thank Ms Irene Nicotra for her help and assistance.

Authors' address

University of Pisa
Dipartimento di Scienze Archeologiche
Via S. Maria 53
56100 Pisa, ITALIA

Bibliography

BARTOLI F., 1996. Paleodieta. I gruppi umani neolitici dell'Italia centro meridionale. In: Forme e tempi della neolitizzazione in Italia meridionale e in Sicilia. Atti del Seminario internazionale di Rossano Calabro, Aprile-Maggio 1994, vol. 2: 499-507.

BARTOLI F., MALLEGNI F., BERTOLDI F., 2001. Analisi paleobiologica e paleonutrizionale dei resti scheletrici rinvenuti nella grotta Continenza. Atti del II Convegno "Il Fucino e le aree limitrofe nell'antichità", Celano, 1999.

BISEL S.C., 1980. A pilot study in aspects of human nutrition in the ancient eastern Mediterranean, with particular attention to trace mineral in several populations from different time periods. PhD dissertation, Washington.

CECCANTI B., 1994. Alterazioni diagenetiche dei reperti ossei nel terreno. In: F. Mallegni, M. Rubini, Recupero dei materiali scheletrici umani in archeologia, a cura di, Roma: 193-222.

FIDANZA F., LIGUORI G., 1988. Nutrizione umana. Idelson, Napoli.

FORNACIARI G., MALLEGNI F., 1987. Paleonutritional studies on skeletal remains of ancient populations from the Mediterranean area: an attempt to interpretation. Antropologischer Anzeiger 45: 361-370.

GILBERT R.J., 1985. Stress, Paleonutrition and Trace Elements. In: Gilbert R.J., Mielke J.H., The Analysis of Prehistoric Diets, a cura di, New York: 339-358.

GRIFONI CREMONESI R., 1998. Alcune osservazioni sul rituale funerario nel Paleolitico superiore della Grotta Continenza. Riv. Sc. Preist., 49: 395-410.

KLEPINGER L.L., 1984. Nutritional assessment from bone. Annual Review of Anthropology, 13: 73-96.

LAMBERT J.B., SZPUNAR C.B., BUIKSTRA J.E., 1979. Chemical analysis of excavated human bone from Middle and Late Woodland Sites. Archaeometry, 21: 115-129.

RICHARDS M. P., PETTITT P. B., STINER M. C., TRINKAUS E., 2001. Stable isotope evidence for increasing dietary breadth in the European mid-Upper paleolithic.PNAS, 98, 11: 6528-6532.

SANDFORD M.K. 1992. A reconsideration of trace element analysis in prehistoric bone. In: Saunders S. e Katzenberg M., a cura di, Skeletal Biology of Past Peoples: research methods: 79-103.Wiley-Liss, New York.

Table 4. Average values of Sr/Ca and Zn/Ca in Iron age samples

	Sr/Ca	Zn/Ca	Number of individuals
Misincinis (UD) IXth-IIIrd cent. BC	0,69±0,20	0,26±0,28	40
Le Ripaie (PI) IX-VIIIth cent. BC	0,75±0,26	0,39±0,21	26
Pontecagnano (SA) IXth cent. BC	0,49±0,05	0,23±0,06	36
Este (PD) VIIIth-IIIrd cent. BC	0,57±0,14	0,27±0,09	23
Ardea (Roma) VIII-VIth cent. BC	0,79±0,16	0,55±0,19	24
Pontecagnano VIIIth cent. BC	0,59±0,11	0,27±0,06	12
Ferrone (Roma) VII-VIth cent. BC	0,61±0,18	0,58±0,25	27
Alfedena (AQ) VIIth-IIIrd cent. BC	0,54±0,44	0,36±0,35	26
Bazzano (AQ) VIIth-IIIrd cent. BC	0,62±0,16	0,32±0,42	42
Pontecagnano (SA) VII-VIth cent. BC	0,93±0,28	0,43±0,14	16
Siracusa (SR) VII-VIth cent. BC	0,80±0,64	0,44±0,19	15
Athens-Greece	0,78±0,18	0,48±0,09	15

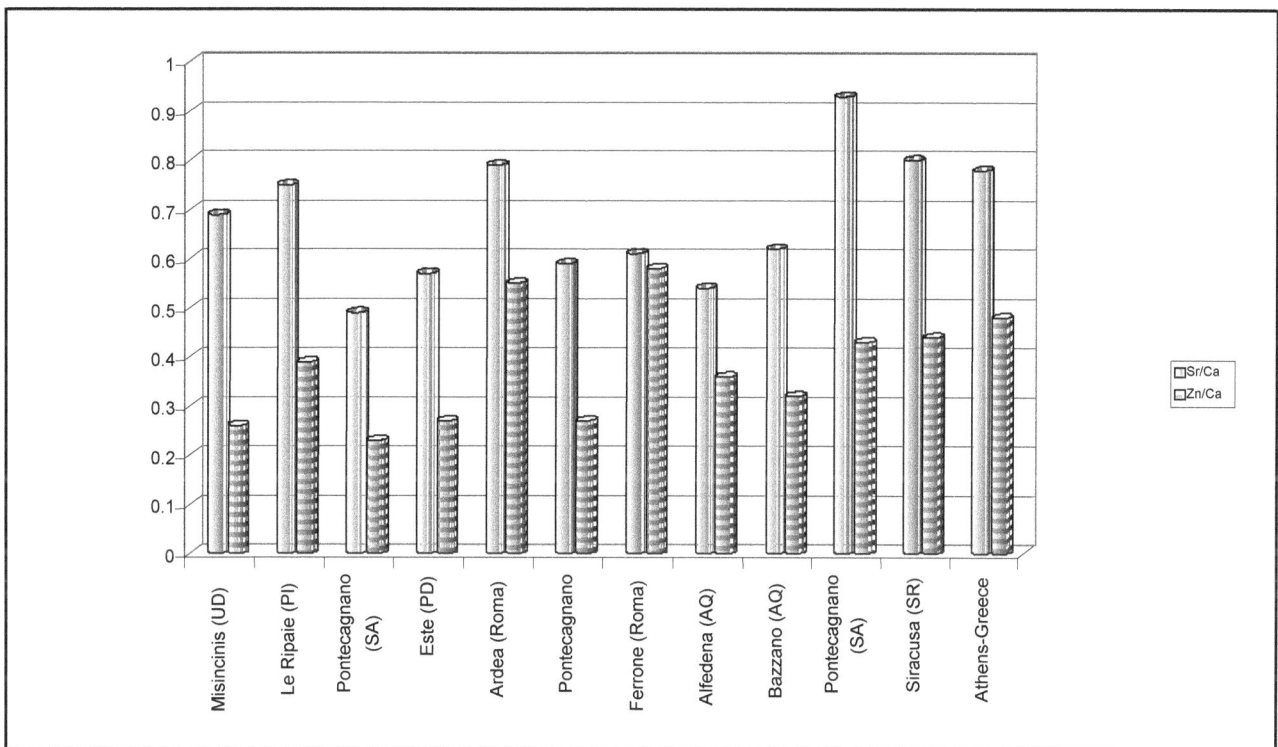

Graph 4. Iron age samples (Sr/Ca and Zn /Ca)

SCHOENINGER M.J., 1980. Changes in human subsistence activities from the Middle East. PhD dissertation, University of Michigan.

SCHOENINGER M.J., 1982. Diet and evolution of modern human form in the Middle East. Amer. Journ. of Phys. Anthrop., 58: 37-52.

SCHOENINGER M.J., 1981. The agricultural "revolution": its effects on human diet in prehistoric Iran and Israel. Paléorient, 7: 73-91.

SILLEN A., 1981 a. Post-depositional changes in Natufian and Aurignacian faunal bones from Hayonim Cave. Paléorient, 7: 81-85.

SILLEN A., 1981 b. Strontium and diet at Hayonim Cave. Amer. Jour. of Phys. Anthrop., 56: 131-138.

SILLEN A., 1984. Dietary changes in the Epi-paleolithic and Neolithic of the Levant: the Sr/Ca evidence. Paléorient, 10: 149-155.

SILLEN A., Kavanagh M., 1982. Strontium and Paleodietary research: a review. Yearbook of Physical Anthropology, 25: 67-90.

TOOTS H., VOORHIES M.R., 1965 - Strontium in fossil bones and the reconstruction of food chains. Science, 149: 854-855.

TUREKIAN K.K., KULP J.L., 1956 - Strontium content of human bones. Science, 124: 405-407.

UNDERWOOD E.J., 1977 - Trace elements in human and animal nutrition. New York.

LE CERF ELAPHE DANS LES SITES DU PALEOLITHIQUE MOYEN DU SUD-EST DE LA FRANCE ET DE LIGURIE. INTERETS BIOSTRATIGRAPHIQUE, ENVIRONNEMENTAL ET TAPHONOMIQUE

Patricia VALENSI, Eleni PSATHI & Frédéric LACOMBAT

Résumé : Les restes de Cervus elaphus de différents sites pléistocènes des Alpes-Maritimes et de Ligurie (Lazaret, Madonna dell'Arma, Fate, Manie, Santa Lucia Superiore) ont été étudiés, sous un angle paléontologique et taphonomique. L'analyse morphométrique des dents met en évidence des variations liées principalement à des changements climatiques et dans un moindre degré à la position chronologique de chaque population. Les données taphonomiques indiquent que le Cerf a toujours été systématiquement chassé et apporté entier dans chaque site en question et ensuite intensivement exploité avec les mêmes méthodes par les hommes préhistoriques.

Abstract: Cervus elaphus remains of different Pleistocene sites of Alpes-Maritimes and Liguria (Lazaret, Madonna dell'Arma, Fate, Manie, Santa Lucia Superiore) have been studied under a palaeontological and taphonomic point of view. The morphological and biometric analysis of teeth remains reveals variations related primarily to climatic changes and secondary to biological evolution. The taphonomic data suggest a systematic human hunting of red deer and transport of the entire carcass of these animals to the camps. Finally, red deer had been widely utilized by humans, using similar butchery procedures between the different sites.

1. INTRODUCTION

La région des Alpes-Maritimes et de Ligurie renferme de nombreux gisements préhistoriques datés de la fin du Pléistocène moyen (stade isotopique 6) et du début du Pléistocène supérieur (stades isotopiques 5, 4 et début du 3). Au cours de ces périodes, *Cervus elaphus* représente le plus souvent l'espèce la plus abondante des assemblages fauniques de la région. L'analyse paléontologique détaillée de cet animal a été réalisée sous diverses approches, notre objectif étant de rechercher certaines caractéristiques morphologiques et/ou métriques de chaque population fossile étudiée et de suivre l'évolution de l'espèce à travers la transition Pléistocène moyen – Pléistocène supérieur, dans un contexte géographique restreint. L'étude taphonomique est également abordée ; elle permet d'affiner l'analyse environnementale et d'apporter des informations sur le mode de vie des hommes préhistoriques concernés (derniers Anténéandertaliens et Néandertaliens).

Notre travail s'appuie principalement sur l'étude paléontologique de grandes collections provenant de quatre gisements importants (Figure 1) : la grotte du Lazaret (Nice, France), Madonna dell'Arma (San Remo, Italie), la Caverna delle Fate et Arma delle Manie (Finale Ligure, Italie). Des données sur l'étude de la faune de la grotte de Santa Lucia Superiore (Toirano, Italie) y sont également intégrées. Enfin, des comparaisons ont été effectuées avec la grotte du Prince (Vintimille, Italie), qui est un gisement de référence pour la région et qui renferme une séquence stratigraphique couvrant les stades isotopiques 6 à 2 (Del Lucchese et al., 1985) et le gisement de Combe-Grenal en Aquitaine qui a livré une population

de cerfs de petite taille *Cervus simplicidens* (Guadelli, 1987).

2. PRESENTATION DES GISEMENTS

La grotte du Lazaret renferme, sur 6 mètres d'épaisseur, une industrie de transition entre acheuléen et moustérien associée à une faune abondante du Pléistocène moyen récent (stade isotopique 6). Des datations sur émail de Cerf par la méthode U-Th/ESR combinée ont donné un âge de 130 000 ans +- 15 000 ans pour l'ensemble supérieur CIII (unités A-D) et 170 000 ans +- 20 000 ans pour l'ensemble inférieur CII (unités E et inférieures) (Michel, 1995).

Madonna dell'Arma comprend un remplissage en grotte d'environ 7 mètres d'épaisseur, attribué au stade isotopique 5. Les dépôts archéologiques situés au-dessus d'une plage thyrrénienne renferment une industrie moustérienne à débitage essentiellement Levallois associée à une faune tempérée.

A la Caverna delle Fate, différents planchers stalagmitiques alternent avec des strates sablo-argileuses riches en matériel archéologique. Le plancher IV a été daté par la méthode ESR d'environ 60 000 ans (Falguères et *alii*, 1990). Au-dessous de ce dernier plancher, les niveaux archéologiques fouillés ont livré une industrie lithique moustérienne et une faune tempérée très diversifiée.

Les niveaux paléolithiques de l'Arma delle Manie, situés au dessus d'un plancher stalagmitique daté d'environ 90 000 ans (Mehidi N., comm. pers.), sont attribués aux stades

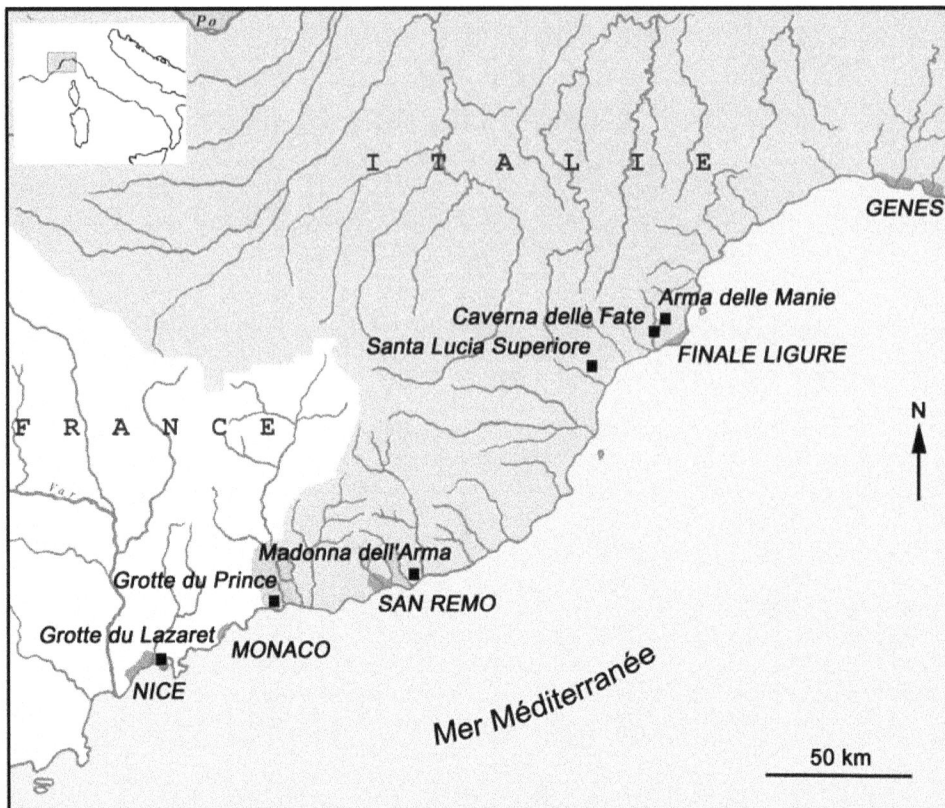

Figure 1 : Situation géographique des gisements

isotopiques 4 et 3 . Ils renferment une faune dominée par des cervidés, associée à une industrie moustérienne.

3. CONTEXTE BIOSTRATIGRAPHIQUE ET ENVIRONNEMENTAL (Figure 2 et Tableau 1)

Les faunes du Lazaret, datés de la fin du Pléistocène moyen (stade isotopique 6), mettent en évidence un climat continental relativement froid et très humide. Les faunes à cachet tempéré restent toutefois les plus abondantes, du fait de la position géographique du site jouant le rôle d'une zone refuge (Valensi & Abbassi, 1998). La population de Cerf élaphe y est très importante et domine largement le reste des grands herbivores sur toute la séquence stratigraphique (Valensi, 2000).

Pendant le stade isotopique 5, la période tempérée est marquée dans la région par des dépôts de la mer thyrrénienne (Grotte du Prince, Madonna dell'Arma, Plage de la villa Marcella à Nice) (Lumley et al., 2001) ou par des planchers stalagmitiques (Grotte du Lazaret, Arma delle Manie).

Les faunes provenant de la deuxième partie du stade isotopique 5 se rencontrent dans les niveaux moustériens de Madonna dell'Arma (couche IV à I) contemporains des foyers E et D de la grotte du Prince (Del Lucchese et al., 1985). On note à cette époque, la prédominance du Cerf et l'association *Elephas antiquus, Stephanorhinus kirchbergensis* et *Hippopotamus incognitus*. Plus dans l'arrière pays, les gisements de Fate et de Manie (couche

VII) renferment des dépôts datés de la fin du stade 5 et du début du stade 4, riches en faunes tempérées forestières : cerfs, chevreuils, sangliers abondants. Les carnivores sont très diversifiés. Dans la grotte de Fate, on note en particulier les dernières apparitions d'*Ursus thibetanus* et de *Hyena prisca*.

Durant le stade isotopique 4, les cervidés restent toujours abondants. Le refroidissement climatique entraîne la diminution progressive des espèces tempérées au profit des éléments caractéristiques d'un environnement plus ouvert sous climat plus froid. Le Bouquetin, présent dans la période précédente, devient plus abondant, accompagné du Chamois et du Cheval. Ces changements fauniques s'observent dans le foyer C de la grotte du Prince et dans les couches VI à IV de Manie.

Le foyer vert (ou couches 5/6) de la grotte du Prince, attribué à la petite amélioration climatique du début du stade 3, semble être contemporain de la couche B2 de Santa Lucia Superiore et de la couche III de Manie.

Le niveau supérieur B1 de Santa Lucia Superiore, plus froid, peut être rapproché de la base du foyer B de la grotte du Prince et probablement de la couche II de Manie.

En résumé, dans les régions provençale et ligure, les stades isotopiques 6 et 4, sont peu marqués par le refroidissement climatique. La forte représentation des faunes à cachet tempéré durant ces périodes est liée à la position géographique des sites, entre le littoral méditerranéen et la barrière alpine.

Tableau 1 : Liste des grands herbivores par site

Espece / site	Lazaret	Madonna	Fate	Manie	St Lucia sup.
Megaloceros giganteus	—				
Cervus elaphus	—				
Dama dama	—				
Rangifer tarandus	—				
Capreolus capreolus	—				
Bos primigenus	—				
Bison priscus	—				— ?
Capra ibex	—				
Rupicapra rupicapra	—		—		—
Sus scrofa	—				
Hippopotamus incognitus		—			
Equus caballus	—				
Stephanorhinus hemitoechus	—				
Stephanorhinus kirchbergensis		—			
Stephanorhinus sp.			—	—	
Coelodonta antiquitatis	—				
Elephas antiquus	—				

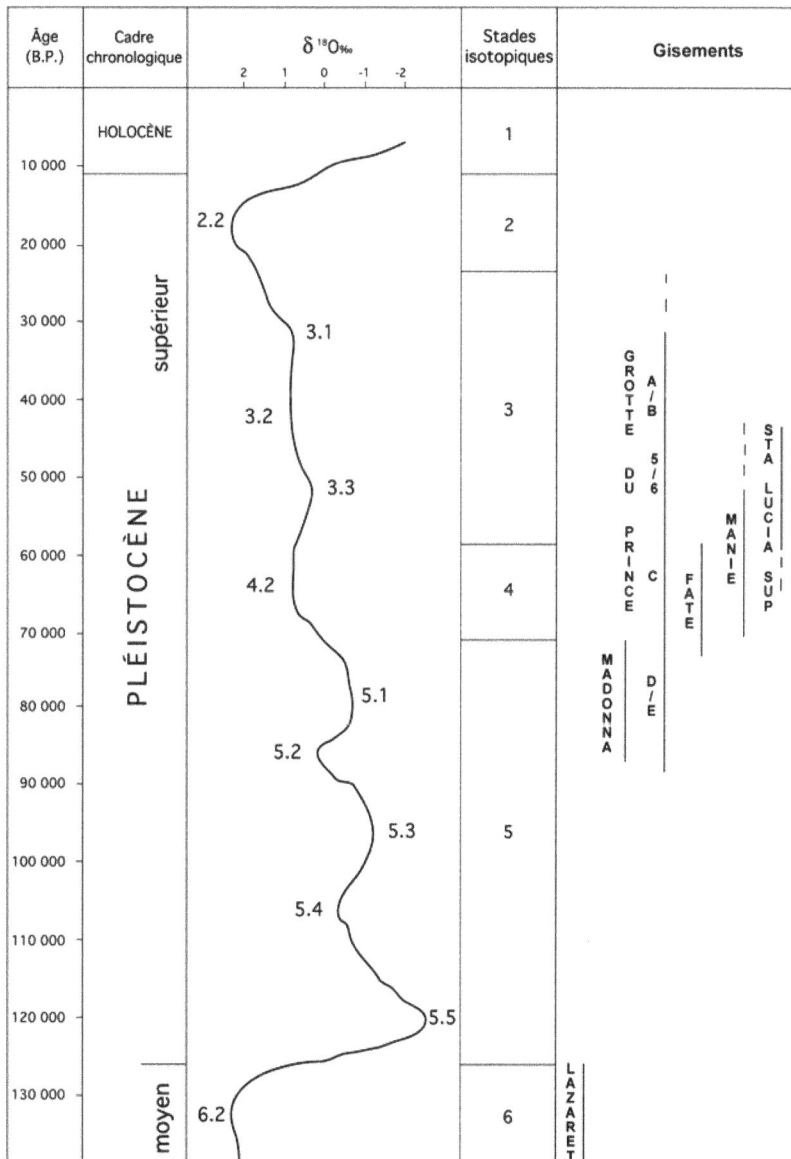

Figure 2 : Datations de quelques gisements de la région du sud-est de la France et de Ligurie

4. LA PRESENCE DU CERF DANS LES SITES ETUDIES

Le Cerf est l'animal le plus abondant dans les gisements préhistoriques depuis le stade isotopique 6 jusqu'au stade 3 (Figure 3). Les variations de représentation de cet animal sont inversement proportionnelles à celles de la deuxième espèce. Au Lazaret, dans les niveaux B et C, le Cerf diminue au profit du Bouquetin, tout en restant l'espèce la plus abondante. A Madonna, la diminution du Cerf se fait au profit de l'Aurochs. Le choix humain joue un rôle important dans la représentation des espèces et surtout pour ces périodes où la chasse est souvent sélective. Néanmoins, ces variations sont d'abord liées à des fluctuations climatiques. Certains niveaux du Lazaret (B et C) et de Manie (couche IV), qui sont plus riches en Bouquetin, correspondent à un climat plus frais, comme il a été démontré par ailleurs (Valensi & Abbassi, 1998). A Santa Lucia Superiore, l'effectif est faible mais la prédominance du Bouquetin pourrait également s'expliquer par la position géographique du gisement, situé plus en retrait du littoral.

5. ETUDE SYSTEMATIQUE- MORPHOLOGIE ET BIOMETRIE

Des Analyses de Correspondances Multiples (ACM) ont été établies à partir de divers critères morphologiques relevés sur les dents jugales et dont certains avaient été proposés par Laquay (1981) puis par Guadelli (1987) lors de son étude sur le *C. simplicidens*. Chaque critère morphologique relevé a ainsi été quantifié et codifié pour chaque dent et pour tous les gisements concernés. De nombreuses données sont actuellement en cours d'exploitation. Nous présenterons ici quelques résultats préliminaires établis sur la morphologie des dents inférieures.

L'analyse de correspondances multiples (Figure 4) établie en fonction des critères morphologiques de la M1 inférieure

permet de faire des regroupements, d'une part sur les critères morphologiques et d'autre part, sur les sites. En utilisant par exemple les critères sur l'ectostylide (développement, morphologie) et ceux sur le pli antérieur (développement, morphologie), il ressort de l'analyse que l'ectostylide est généralement de morphologie simple, lorsque celui-ci est peu développé ; et que dans ce cas, le pli antérieur est souvent faiblement développé. D'autre part, en prenant en compte les sites (Figure 5), ces analyses hiérarchiques montrent que les populations de Manie et de Fate sont très proches par leur morphologie dentaire. La population de Madonna se détache des précédentes (ectostylide simple et pli antérieur faiblement développé). Celle du Lazaret présente de manière générale des ectostylides de forme simple mais toujours développés.

Les résultats de l'étude biométrique présentés sous la forme d'un diagramme des rapports (Figure 6) mettent en évidence des populations de Cerf de petite taille (Madonna dell'Arma ; *C. simplicidens* de Combe-Grenal inférieur (Cgi) par rapport aux populations des autres sites (Lazaret, Manie, Fate et Combe-Grenal supérieur (Cgs). Ces différences de taille sont plus marquées sur les dents jugales inférieures que sur les dents supérieures. On notera également que les populations de petite taille présentent en général des dents de plus forte robustesse.

L'analyse hiérarchique, établie à partir de la biométrie de la M1 inférieure, confirme ces observations (Figure 7). Sur le graphe 7B, en procédant à une coupure (en « 1 »), il est possible de subdiviser ces populations en deux groupes, basés sur la taille. Le premier groupe réunit la population de Cerf de Madonna et le *Cervus simplicidens* de Combe-Grenal, attribués à la fin du stade isotopique 5. Le deuxième groupe rassemble les populations de plus grande taille : Fate, Manie et les niveaux supérieurs de Combe-Grenal, contemporains des stades isotopiques 4 et début du stade 3 ainsi que la population du Lazaret, datée du stade 6. Orgnac 3, gisement

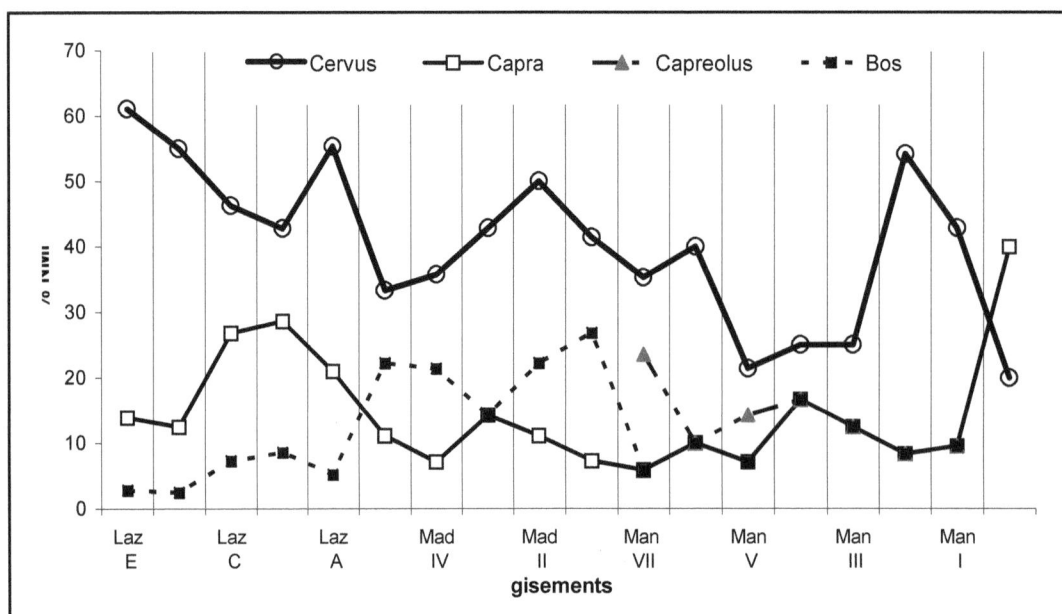

Figure 3 : Abondance du Cerf (en NMI) par rapport aux autres taxons principaux.

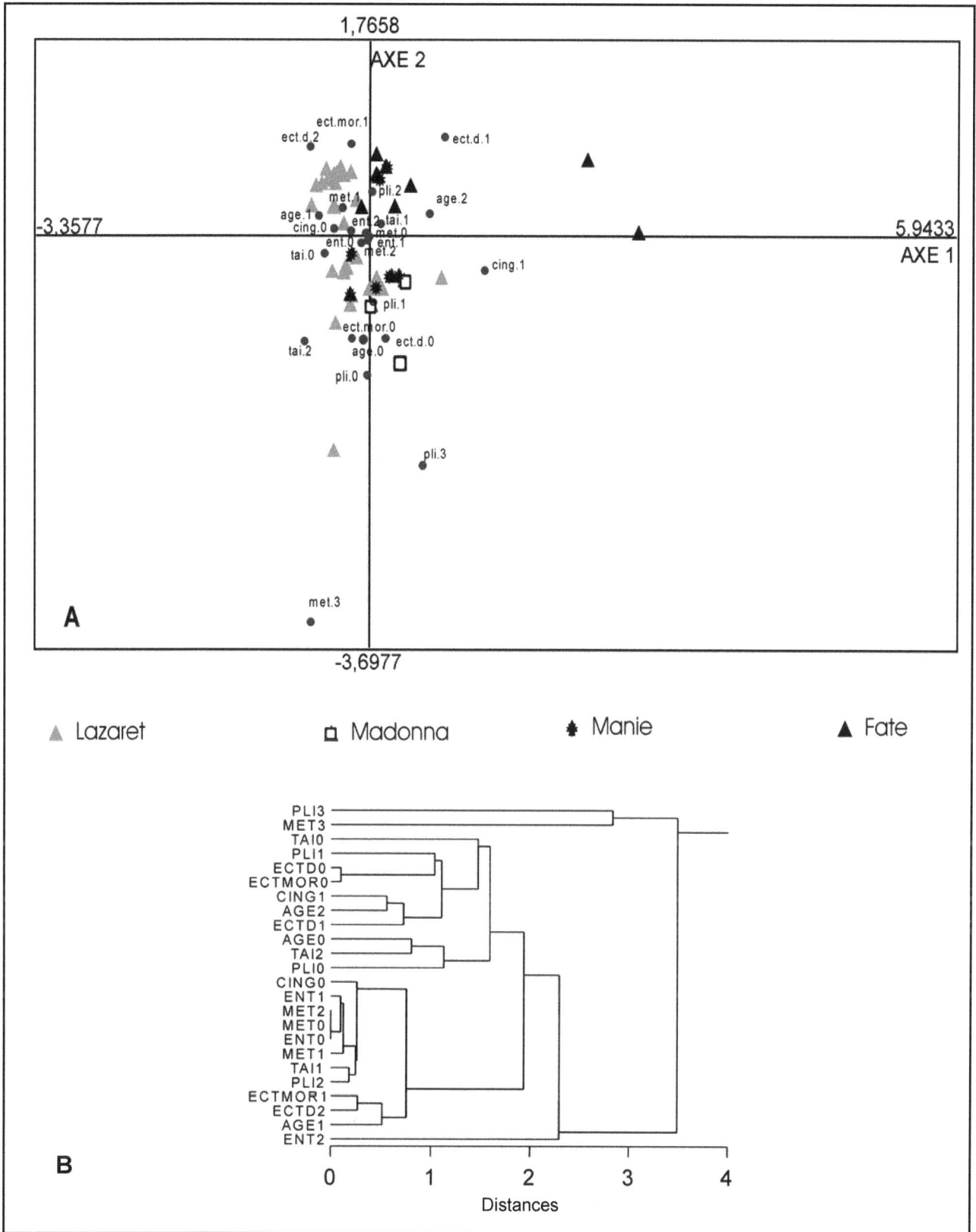

Figure 4 : Analyse de correspondances multiples (ACM) (A) et analyse hiérarchique
(B) en fonction des critères morphologiques de la M1 inférieure.

ardéchois (Sud-Est de la France) daté du stade isotopique 9, présente une population de petite taille, proche du premier groupe. Une coupure plus précise, procédée en « 2 » sur l'analyse hiérarchique, met en évidence quatre populations de Cerf. Ces mêmes groupements se lisent également sur

l'analyse à composantes multiples, avec l'axe 1 pouvant se définir comme l'axe de « taille » et l'axe 2, probablement comme l'axe de « robustesse des dents ». Ainsi, sur la figure 7A, l'axe 1 permet d'individualiser des populations de petite taille de période tempérée et des populations de plus grande

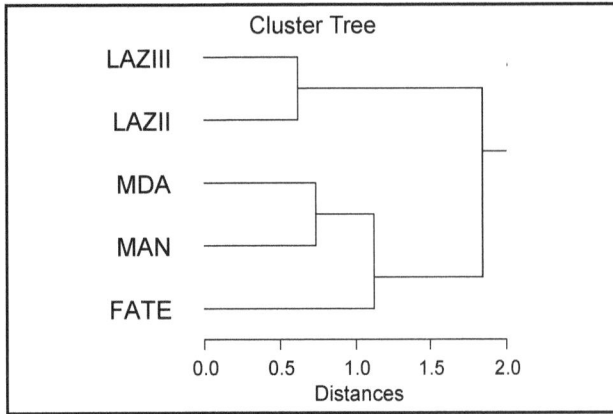

Figure 5 : Analyse hiérarchique des populations de Cerf en fonction de la morphologie de la P4 inférieure.

taille, de climat plus frais. L'axe 2 sépare les populations du Pléistocène supérieur de celles du Pléistocène moyen.

6. DONNEES TAPHONOMIQUES

L'étude taphonomique a été conduite sur les sites du Lazaret, Manie, Fate et Madonna dell'Arma. La fonction principale

de ces gisements correspond à des habitats en grotte, de plus ou moins longue durée en fonction des niveaux archéologiques de chaque site. Entre les différents gisements, les modes d'acquisition et d'exploitation du cerf présentent des similitudes mais aussi quelques spécificités.

Les profils de mortalité des quatre populations de Cerf étudiées (Figure 8) montrent une prépondérance d'individus adultes dans les sites du Lazaret, de Madonna et de Manie, alors qu'à Fate les faons sont les plus abondants. Ces profils de mortalité caractérisent en général des courbes de prédation. L'étude de l'éruption et de l'usure des dents des jeunes, suggère à Fate, une occupation essentiellement au début de l'automne, mais qui pouvait s'étaler tout le long de l'année. Au Lazaret, l'étude de la saisonnalité montre une occupation de longue durée dans les niveaux supérieurs, et de période plus courte (automne et hiver) dans les niveaux inférieurs.

L'observation des différents éléments squelettiques nous indique que tous les os du squelette de Cerf sont présents sur les sites d'habitat, mais avec toutefois certains déficits en éléments osseux (Figure 9). D'une manière générale, les os de forte densité, donc les plus résistants à une destruction post-dépositionelle, sont les mieux représentés. D'autre part, les os à plus grand intérêt nutritif ne sont pas obligatoirement

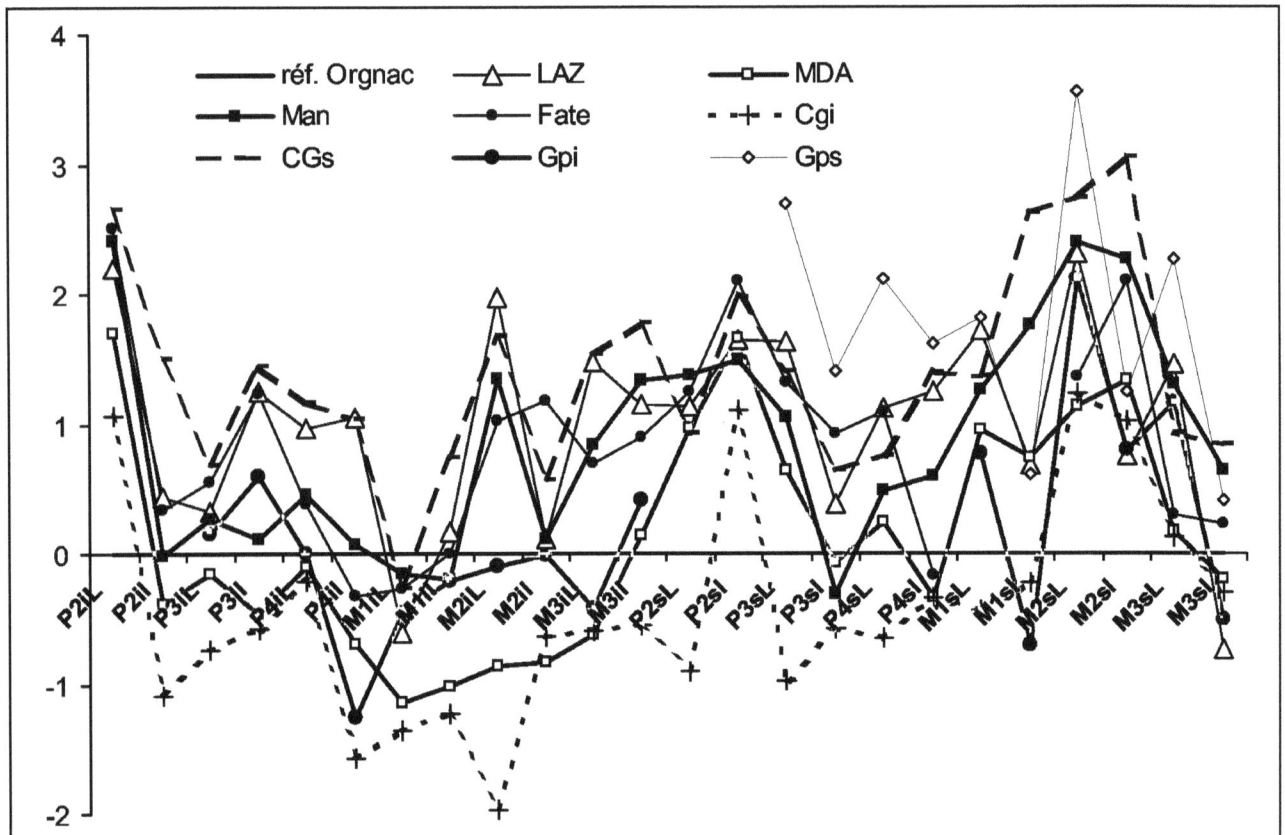

Figure 6 : Diagramme des rapports des dents inférieures et supérieures de *Cervus elaphus*.. Référenciel : Orgnac 3 (Aouraghe, 1992). Comparaisons avec la Grotte du Prince (Arellano-Moullé, 1997-98) et Combe-Grenal (Guadelli, 1987).

L'axe 1, qui est l'axe principal, peut se définir également comme un axe climatique et l'axe 2, comme un axe chronologique. Ainsi, les variations de la taille de nos populations de cerfs sont étroitement liées aux fluctuations climatiques et dans une moindre mesure sont d'ordre chronologique. En résumé, sous un climat tempéré, les populations de cerfs de Provence et de Ligurie sont de plus petite taille avec une morphologie dentaire plus simple et une robustesse des molaires plus grande que celles de climat plus frais.

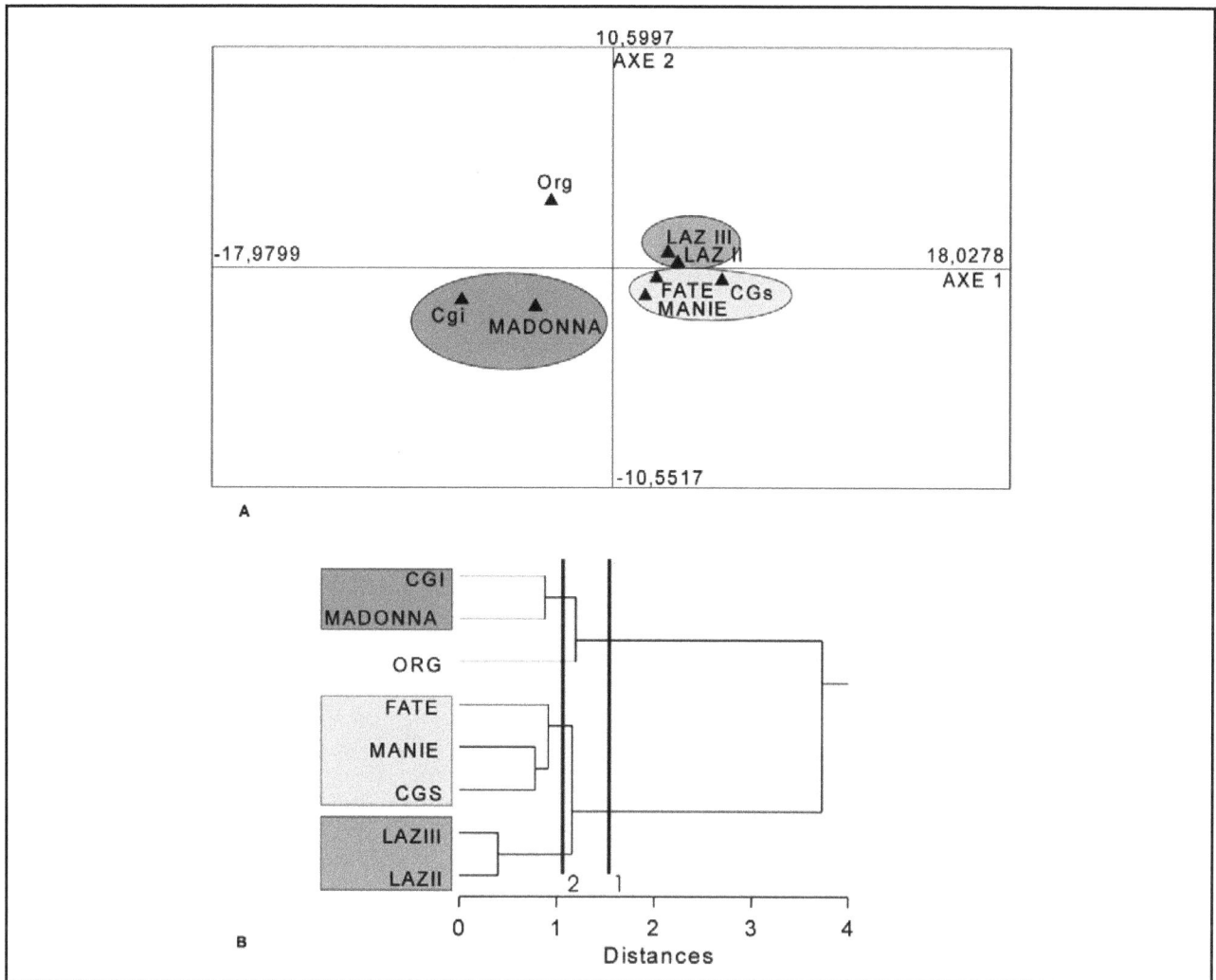

Figure 7 : Analyse à composantes principales (ACP) (A) et analyse hiérarchique
(B) effectuées à partir de la biométrie (DMD max. et DVL max.) des jugales supérieures et inférieures.

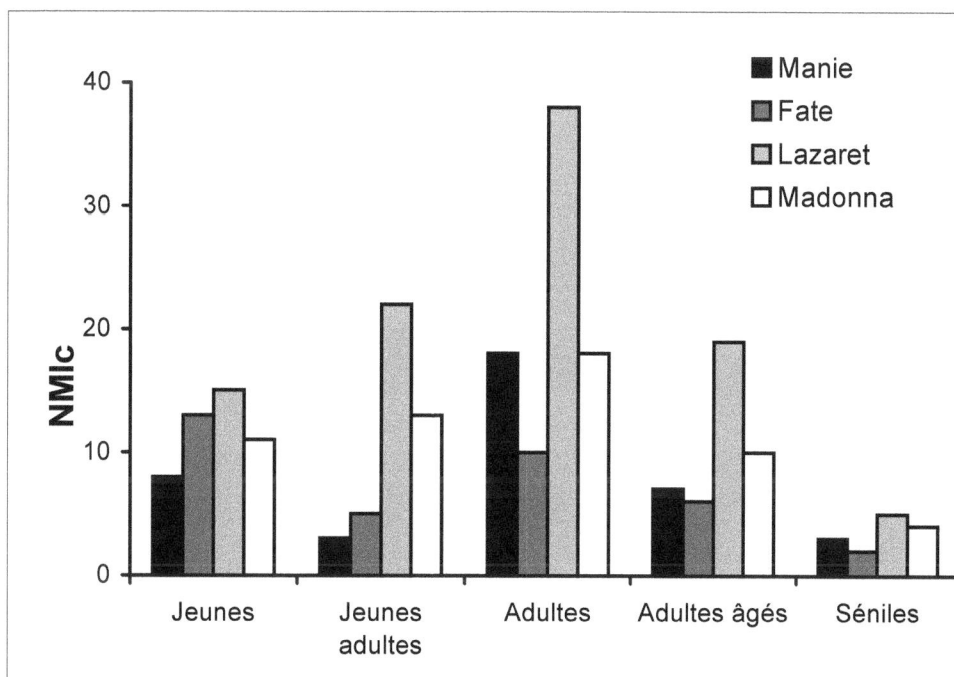

Figure 8 : Profil de mortalité des différentes populations de Cerf (en NMI de combinaison).

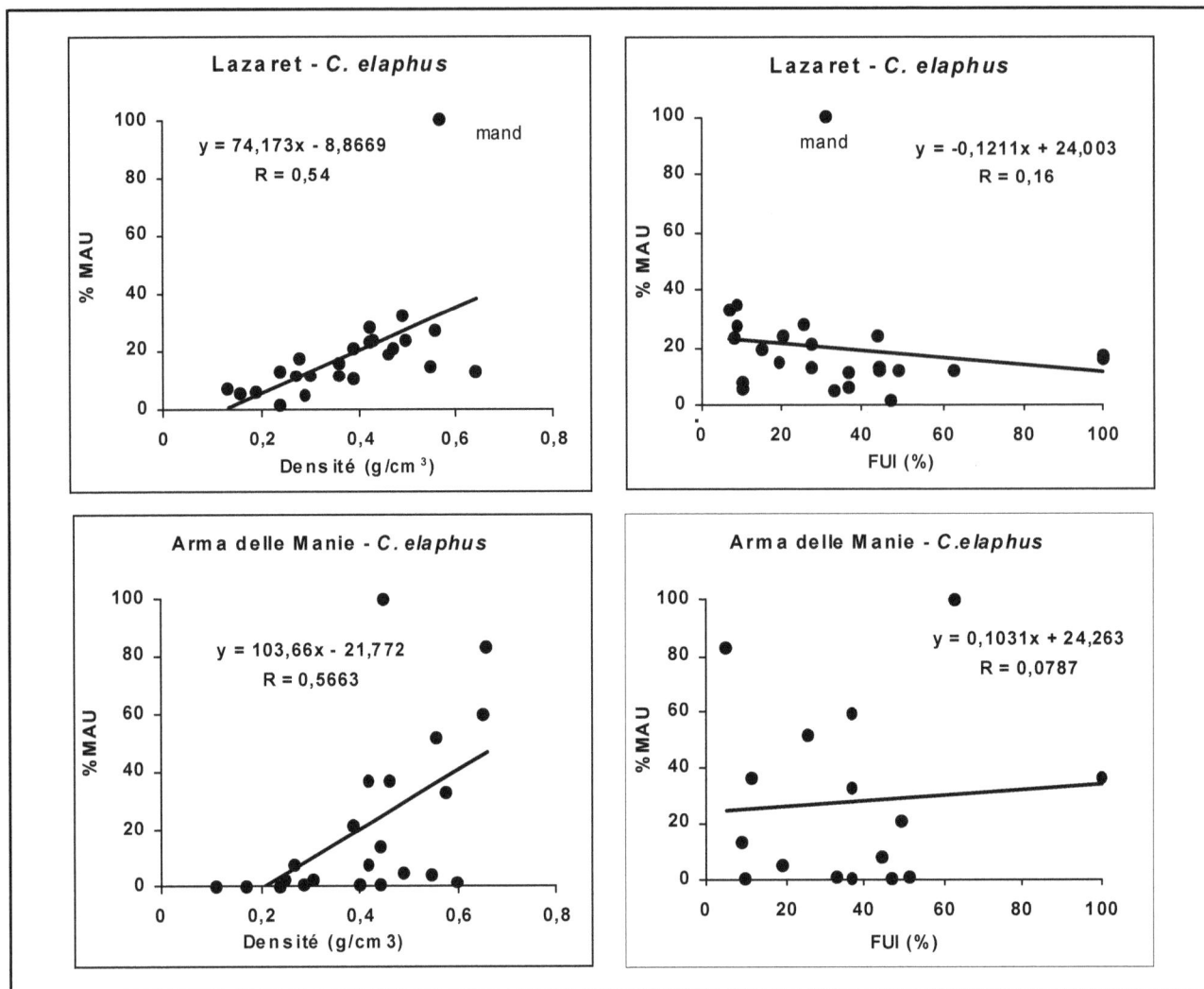

Figure 9 : Relation entre l'abondance des éléments anatomiques (en MAU%, Binford, 1984) de Cerf en fonction de leur densité (Lyman, 1984) et de leur indice d'utilité nutritive (FUI% de Metcalfe & Jone, 1988). Comparaison entre le Lazaret et Manie.

les plus abondants. Les déficits en parties squelettiques sont liés à la conservation différentielle mais également à un traitement différentiel des carcasses par l'Homme. Ces résultats ont été observés au Lazaret (Valensi & Michel, 1996 ; Valensi, 2000) et dans la grotte de Manie (Psathi, en prép.). A Manie par exemple, le déficit en parties spongieuses (extrémités articulaires) pourrait être lié, non seulement à la conservation différentielle, mais aussi aux pratiques culinaires (fracturation des articulations, cuisson, bouillon) ou au mode de traitement des carcasses (éléments de combustion).

Les carcasses de Cerf étaient apportées entières sur le site d'habitat, comme en témoigne la représentation des parties squelettiques. Cette conviction est renforcée par l'examen de la surface osseuse des restes de cet animal. Les marques de percussion (encoches, points d'impact, contre-coups) et les nombreuses stries de découpe permettent de retracer les différentes étapes de traitement des carcasses (écorchement, désarticulation, décharnement, extraction de la moelle osseuse) (Valensi, 2000). Dans tous les sites étudiés, nous avons également noté l'existence de certaines pièces ayant

servi occasionnellement d'outils non aménagés : il s'agit de retouchoirs (ou percuteurs tendres) et de pièces présentant un poli d'utilisation.

7. DISCUSSIONS

En Provence et en Ligurie, le Cerf correspond à l'espèce la plus abondante des assemblages fauniques de la fin du Pléistocène moyen et du début du Pléistocène supérieur. L'examen morphologique et biométrique des dents a révélé l'existence de différentes populations fossiles. La grotte du Lazaret et les grottes de Fate et de Manie renferment des cerfs élaphes de grande taille aux dents morphologiquement complexes. La grotte de Madonna dell'Arma renferme un cerf de petite taille aux dents morphologiquement plus simples et plus robustes, proches de *Cervus simplicidens*.

C'est en 1971 que F. Prat et C. Suire signalent l'existence de deux populations distinctes de cerfs dans le site de Combe-Grenal. Ces deux types ne se rencontrent pas dans les mêmes contextes : les niveaux inférieurs, tempérés, (« Würm I ») renfermant un cerf de petite taille et les dépôts supérieurs, de climat plus rigoureux (Würm II) mettant en évidence un cerf de plus grande taille. Ces auteurs proposent alors d'utiliser ces deux populations de cerfs comme éléments de datation pour le début du Pléistocène supérieur (Prat et Suire, 1971). En 1987, J.-L. Guadelli crée la sous-espèce *C.e.simplicidens* pour le cerf de petite taille de Combe-Grenal, qu'il élève par la suite au rang d'espèce *C. simplicidens* (Guadelli, 1987, 1996). Cette espèce est ensuite reconnue dans divers dépôts du stade isotopique 5 du Sud-Ouest et de l'Ouest de la France, ainsi qu'en Espagne. Citons par exemple la couche 10 d'Artenac en Charente (Delagnes et al., 1999) et des dépôts de l'Aven de Saint-Projet et de Rochelot, en Poitou-Charente (Tournepiche, 1996). L'espèce *C. simplicidens* est par la suite reconnue dans les dépôts du stade 6 de Combe-Grenal, associée au *Cervus elaphus* de grande taille (Delpech et Prat, 1995). En Provence, la présence du Cerf de petite taille est attestée pour la première fois dans les niveaux moustériens de la grotte de l'Adaouste, par Crégut-Bonnoure qui le rapproche par ses faibles dimensions du *C. simplicidens* (Defleur et al., 1994). Enfin, en Ligurie, l'étude paléontologique que nous avons conduite sur le cerf de Madonna dell'Arma met également en évidence des affinités morphologiques et biométriques avec l'espèce de petite taille *C. simplicidens*.

Les variations de taille et de morphologie dentaire sont essentiellement d'ordre climatique. Il pourrait s'agir de modifications morphométriques liées à la loi de Bergman et jouant sur une même population de cerfs. Certains auteurs (Sacchi & Testard, 1971, p. 393) rappellent toutefois que cette loi écologique a un rôle négligeable sur les populations de Cerf commun européen. Les données paléontologiques actuelles semblent en fait confirmer l'existence des deux populations de cerfs au Pléistocène moyen récent et au Pléistocène supérieur, l'une de grande taille *Cervus elaphus* et l'autre de petite taille dont les dents sont à ornementation plus simple, et acceptée par certains auteurs comme une espèce différente *Cervus simplicidens*. Il semble qu'au cours des périodes tempérées de l'Europe occidentale, l'espèce de petite taille se développe au profit du grand Cerf et en revanche, durant les périodes froides, le Cerf de grande taille à ornementation dentaire complexe devient l'espèce prépondérante. Ces données semblent se confirmer grâce à la présence des deux populations dans les mêmes niveaux du Pléistocène moyen de Combe Grenal (Delpech et Prat, 1995).

Chronologiquement, et d'après les données actuelles, la population de Cerfs de petite taille est présente au moins dès la fin du Pléistocène moyen (stade isotopique 6) et se développe durant les périodes tempérées du Pléistocène supérieur, au cours des stades 5 voire 3. Du point de vue géographique, cette population est décrite ici pour la première fois en Ligurie (Madonna dell'Arma).

L'étude paléontologique que nous avons réalisée devra être complétée par un examen plus détaillé de la morphologie des dents et des os du post-crânien. Dans l'état actuel des données, nous attribuerons le petit Cerf de Madonna dell'Arma à *Cervus elaphus aff. simplicidens*. La présence de ce Cerf de petite taille peut être utilisée comme un bon indicateur environnemental et indirectement chronologique.

Le cerf qui est abondant dans les assemblages fauniques découverts en contexte préhistorique, permet également d'apporter des informations essentielles sur le mode de vie des hommes préhistoriques des régions provençale et ligure.

Abréviations

Abréviations des critères morphologiques. tai :taille, cing : développement du cingulum, ect.mor. : morphologie de l'ectostylide, ect.d. : développement de l'ectostylide, pli : développement du pli antérieur, met. : morphologie du métaconide, ent. : morphologie de l'entostylide. *Codes* : 0 : absent, 1 : peu développé, 2 : très développé. *Abréviations des mesures dentaires*. L ou DMD : longueur ou diamètre mésio-distal, l ou DVL: largeur ou diamètre vestibulo-lingual, i : inférieur, s : supérieur. *Abréviations des sites*. *Lazaret* – Niveaux supérieurs (Laz III) : unités A à D. Niveaux inférieurs (Laz II) : unité E ; *Manie* (Man) ; *Madonna dell'Arma* (Mad) ; Grotte *du Prince* - Niveaux inférieurs (Gpi) : foyers E ,D, C. Niveaux supérieurs (Gps) : foyers A,B. ; *Combe-Grenal* - Niveaux inférieurs (Cgi) : couches 54-50. Niveaux supérieurs (CGs) : couches 35-1 ; *Orgnac 3* (Org)

Remerciements

Les auteurs remercient Henry de Lumley, Annie Echassoux (Lab. dép. de préhistoire du Lazaret), Carlo Tozzi (Université de Pise), Massimo Ricci (Museo civico de San Remo) et Giuseppe Vicino (Museo Archeologico del Finale) qui nous ont confié l'étude du matériel archéologique. Ce travail n'aurait pu être réalisé sans la collaboration de *La Soprintendenza archeologica della Liguria* ; nos remerciements vont en particulier à Mme G. Spadea, la soprintendente, et Mrs. R. Maggi et A. Del Lucchese. Merci également à Khalid El Guennouni pour son aide sur les analyses statistiques.

Adresses des auteurs

Laboratoire départemental de Préhistoire du Lazaret
33bis, Boulevard Franck Pilatte
06300 Nice-FRANCE
valensi@lazaret.unice.fr
epsathi@lazaret.unice.fr
flacombat@lazaret.unice.fr

Bibliographie

AOURAGHE, H., 1992, *Les faunes de grands mammifères du site Pléistocène moyen d'Orgnac 3 (Ardèche, France). Etude paléontologique et palethnographique. Implications paléoécologiques et biostratigraphiques* Thèse Doctorat Muséum National d'Histoire Naturelle, Paris.

ARELLANO-MOULLÉ, A., 1997-1998, Les Cervidés des niveaux moustériens de la Grotte du Prince (Grimaldi, Vintimille, Italie). Etude paléontologique. *Bulletin Musée Anthropologie préhistorique de Monaco* 39, p. 53-58.

BINFORD, L.R., 1984, *Faunal remains from the Klasies River Mouth.* New-York : Academic Press.

DEFLEUR, A., BEZ, J.-F., CREGUT-BONNOURE, E., DESCLAUX, E., ONORATINI, G., RADULESCU, C., THINON, M. & VILETTE, P., 1994, Le niveau moustérien de la grotte de l'Adaouste (Jouques, Bouches-du-Rhône). Approche culturelle et paléoenvironnements. *Bulletin Musée Anthropologie préhistorique de Monaco* 37, p. 11-48.

DELAGNES, A., TOURNEPICHE, J.-F., ARMAND, D., DESCLAUX, E., DIOT, M.-F., FERRIER, C., LE FILLATRE, V. & VANDERMEERSCH, B., 1999, Le gisement Pléistocène moyen et supérieur d'Artenac (Saint-Mary, Charente) : premier bilan interdisciplinaire. *Bulletin de la Société Préhistorique Française* 96, 4, p. 469-496.

DEL LUCCHESE, A., GIACOBINI, G. & VICINO, G., dir., 1985, *L'Uomo di Neandertal in Liguria.* Quaderni della Soprintendenza Archeologica della Liguria, 2. Genova : Tormena Editore.

DELPECH, F. & PRAT, F., 1995, Nouvelles observations sur les faunes acheuléennes de Combe Grenal (Domme, Dordogne). *PALEO* 7, p. 123-137.

GUADELLI, J.-L., 1987, *Contribution à l'étude des zoocénoses préhistoriques en Aquitaine (Würm ancien et interstade würmien).* Thèse Doctorat, Université de Bordeaux I.

GUADELLI, J.-L., 1996, Les cerfs du Würm ancien en Aquitaine. *PALEO* 8, p. 99-108.

FALGUÈRES, C., YOKOYAMA, Y. & BIBRON, R., 1990, Electron Spin Resonance (ESR) Dating of Hominid – Bearing Deposits in the Caverna delle Fate, Ligure, Italy. *Quaternary Research* 34, p.121-128.

LAQUAY, G., 1981, *Recherches sur les faunes du Würm I en Périgord.* Thèse 3° cycle, n° 1596, Université de Bordeaux I.

LUMLEY, H. de, KHATIB, S., ECHASSOUX, A. & TODISCO, D., 2001, Les lignes de rivages quaternaires dans les Alpes-Maritimes et dans la Ligurie italienne. In *Groupe des Méthodes Pluridisciplinaires contribuant à l'Archéologie, 4ème Colloque d'Archéométrie*, La Rochelle, 24-28 avril 2001, p. 35, résumé.

LYMAN, R.L., 1984, Bone density and differential survivorship of fossil classes. *Journal of Anthropological Archaeology* 3-4, p. 259-299.

METCALFE, D. & JONE, K.T., 1988, A reconsideration of animal body-part utility indices. *American Antiquity* 53, 3, p.486-504.

MICHEL, V., 1995, *Etude des influences des processus de fossilisation sur le fondement de la datation radiométrique. Application à la datation par U-Th et ESR de mâchoires (os, dent) de* Cervus elaphus *des niveaux archéologiques de la grotte du Lazaret.* Thèse Doctorat, Muséum National d'Histoire Naturelle, Paris.

PRAT, F. & SUIRE, C., 1971, Remarques sur les cerfs contemporains des deux premiers stades würmiens. *Bulletin de la Société préhistorique française* 68, 3, C.R.S.M., p. 75-79.

PSATHI, E., en prép., *Les sites de l'Arma delle Manie et de la Caverna delle Fate (Ligurie, Italie). Etude paléontologique, archéozoologique et biostratigraphique de la grande faune.* Thèse Doctorat, Muséum National d'Histoire Naturelle, Paris.

SACCHI, C.F. & TESTARD, P., 1971, *Ecologie animale. Organismes et milieu.* Paris : Doin.

TOURNEPICHE, J.-F., 1996, Les grands Mammifères pléistocènes de Poitou-Charente. *PALEO* 8, p. 109-141.

VALENSI, P. & MICHEL, V., 1996, Taphonomie et fossilisation des restes fauniques de la grotte du Lazaret (Nice, France). In *Taphos 96, II Reunion de Tafonomia y fosilización, Zaragoza.* Diputación de Zaragoza : Fundación Pública de la Excma, p. 401-406.

VALENSI, P. & ABBASSI, M., 1998, Reconstitution de paléoenvironnements quaternaires par l'utilisation de diverses méthodes sur une communauté de mammifères- Application à la grotte du Lazaret. *Quaternaire* 9, 4, p. 291-302.

VALENSI, P., 2000, The Archaeozoology of Lazaret cave (Nice, France). *International Journal of Osteoarchaeology* 10, p. 357-367.

BRONZE AGE AGRICULTURAL IMPACTS
IN THE CENTRAL PART OF THE CARPATHIAN BASIN

Pál SÜMEGI, Imola JUHÁSZ, Elvira BODOR &Sándor GULYÁS

INTRODUCTION

The closure of the Bronze Age witnessed the initiation of some really important events regarding the future history and development of mankind. This was the time when the first horses were tamed and domesticated by humans on the steppes of Eastern Europe. Even though horses were used purely for their meat at the beginning, the situation has dramatically changed by the dawn of the Bronze Age when their traditional role was complemented with cargo carriage, ploughing and pulling the yoke as well as the transportation of troops. The appearance of the new bronze tools which were on one hand harder than those made of copper, could serve several functions and on the other hand also refusible caused such dramatic changes in the lives and development of humans as the domestication of the horse. These technical and social innovations brought forth the birth of long-range intended warfare accompanied with the emergence of a new social class choosing it as a profession. As the military emerged so did its special tools and weapons, e.g. the sword which was clearly and easily distinguishable from those used for hunting. Relying upon the agility and strength of the horse in cargo carriage the newly developed military was enabled to initiate a planned ambush or attack from larger distances. The sudden appearance of this new type of aggressive force has called for various defense strategies and methods such as the construction of fortified settlements surrounded by deep ditches and palisades or a concatenation of these.

One of the most developed regions of Bronze Age Europe was that of the Carpathian Basin (KOVÁCS, 1977), where a large number of settlements bearing different sizes and structures established by several highly developed productive communities could be traced (Fig. 1). The development of the society culminated during the Middle Bronze Age. At the peak of their evolution within the area of the Carpathian basin these Bronze Age communities occupied highly alternating environments ranging from areas along major brooks and creeks, alluvial plains of the rivers, the foothill areas as well as loess ridges establishing a large number of settlements, the so-called tells, which remained populated for hundreds of years. Even in times preceding the Middle Bronze Age, i.e. during the Late Neolithic as a result of social processes deriving from and hiving off the areas of the Middle East certain settlements were transformed into tell settlements in the area of the Great Hungarian Plains. However, the Bronze Age tell settlements are clearly separated both in space and regarding their structural makeup from those of the Late Neolithic. The fortified Bronze Age tell settlements were surrounded by a network of water ditches reaching depths of 3-6 ms in a circle or semi-circle and very often linked to the

beds of active rivers or those flooded during times of major floods only and oxbow lakes (Fig. 2). Although we have traces of circular water ditches surrounding certain settlements in the area of the Carpathian basin from as early as the Neolithic, up to date they have been interpreted to have been used for sacral and not defensive purposes. Besides these ditches, a part of the settlements have been encircled by multiplied wooden palisades of 2-3ms height offering better protection.

According to results of the analyses of Bronze Age cemeteries connected to the tell settlements these fortified tells can be regarded as major centers of the military aristocracy serving as local focal points of power in the gradually developing hierarchical network of settlements. On the other hand they also must have developed into centers of local agricultural production as it has been assumed on the basis of the analysis of a wide range of archeobotanical finds, i.e. remnants of plants and animal bones excavated from the tell profiles (GYULAI, 1993).

The major goals of our work presented herein was to examine on one hand whether these Middle Neolithic tell settlements could truly serve as centers of agricultural production. Furthermore to trace the possible human impact posed by the agricultural production of these Bronze Age communities on the surrounding past environment in the heart of the Carpathian basin, i.e. in the area of present-day Hungary.

METHODS AND TOOLS

Human landscape utilization, the effects of agriculture and the transformation of the vegetation and soil conditions as a result of human activity can ideally be traced via detailed geoarcheological analysis of archeological localities (EDWARDS, 1991) or paleoecological studies on minor lacustrine basins serving as sedimentary traps (WILLIS et al. 1998).

During the course of our investigations we have carried out detailed palynological analysis of sedimentary sequences derived from marshes, infilled riverbeds in the direct neighborhood of as well as water ditches surrounding a number of Bronze Age tells in Hungary (Szászhalombatta - Földvár, Polgár - Ásotthalom, Polgár - Kenderföld, Szakáld – Testhalom). This has been further complimented with the palynological analysis of samples taken from more distant marshlands and oxbow lakes situated in the neighborhood of the tells (Fig. 1) in order to make direct comparison of the results gained with other pollen data as well as the outcome of preceding environmental historical and archeobotanical

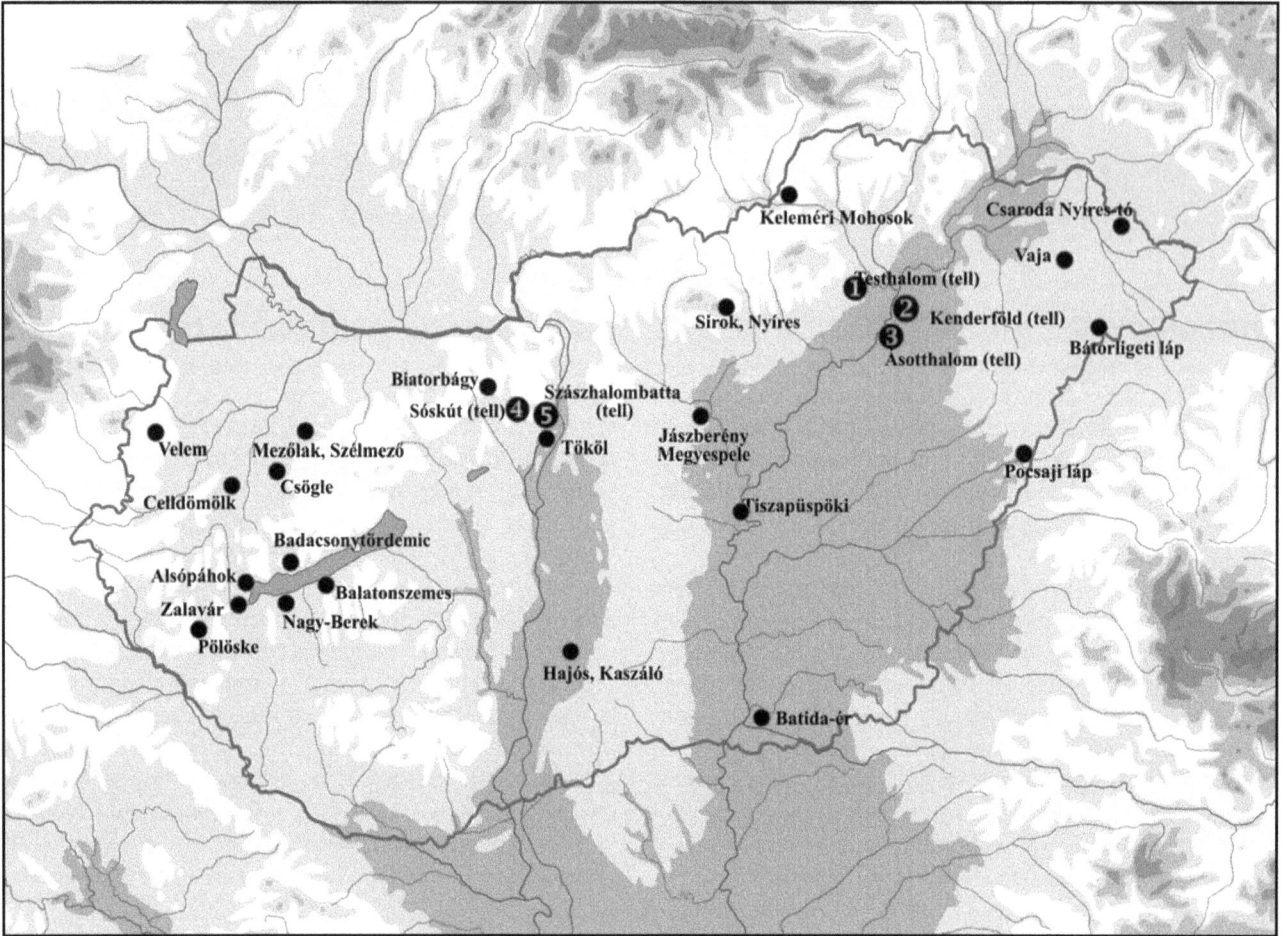

Fig 1. Analysed tells and pollen profiles in Hungary.

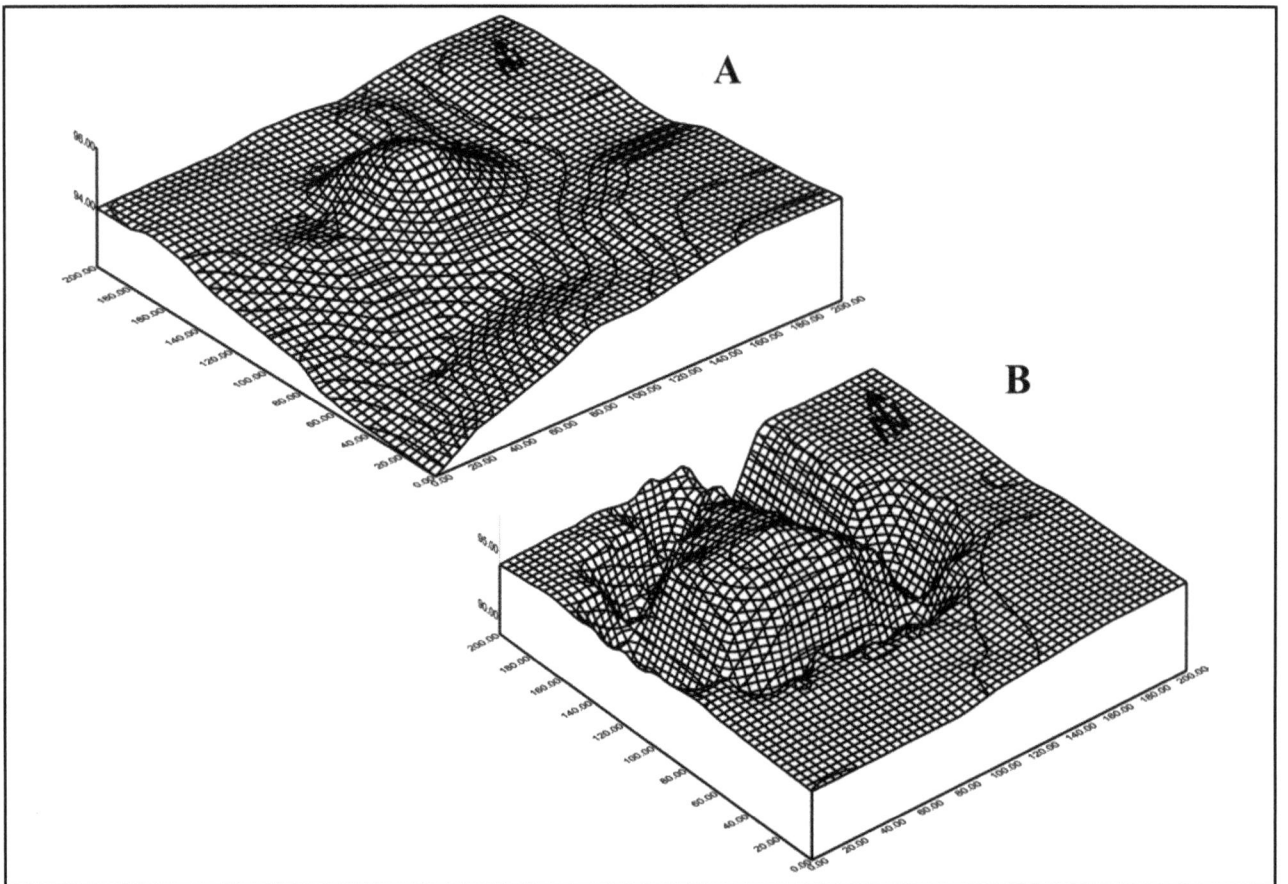

Fig.2. Kenderföld Bronze Age tell at Polgár in Hungary: A = recent stage, B = reconstructed stage.

investigations in the area (WILLIS et al. 1998, SÜMEGI et al. 1998, SÜMEGI, 1999, GARDNER, 1999, GYULAI, 1993, MAGYARI et al. 2002, SÜMEGI-BODOR, 2000, JUHÁSZ, 2002, JUHÁSZ et al. 2002). Data gained from 26 pollen profiles derived from 16 sedimentary basins has been utilized. The material originating from the boreholes and Bronze Age tell profiles was radiocarbon dated in order to enable the direct chronological synchronization of the newly gained data with those on the Middle Bronze Age profiles (RACZKY et al. 1992). The basis of this correlation was the radiocarbon database set up from more than 50 radiocarbon dates gained from material of the tells as well as the approximately 80 dates deriving from the analysis of core material put down into several sedimentary basins. Traces of human activity such as deforestation, the growing of plants, animal husbandry as well as road and trail construction were identified in our pollen data with the help of so called marker species signifying anthropogenic effects (BEHRE, 1988), taking into account their first appearance and spreading in the area (Fig. 3.).

RESULTS

According to our pollen analytical results the prevailing wet climate and the decreasing summer and increasing winter mean temperatures during the Bronze Age must have resulted in the development of balanced temperature conditions in the western, northern and eastern parts of the Carpathian basin. Meanwhile in the central parts there has been an increase in the amount of precipitation compared to the previous pollen zones, although no signs indicating temperature changes could have been found.

There has been a decrease in the number of fir pollens indicating inwash from distant areas in the pollen profiles deriving from the western parts of the basin. This must have resulted in the development of oak forests mixed with beeches and hornbeam. However, this has been accompanied by the appearance of the pollen of grasses and weeds growing alongside of arables, pastures and dirt roads in the pollen profiles. It must be noted that interestingly the increase of hazel pollen (*Corylus*) in several profiles coincided with the closing of the gallery of forests and the appearance and initiating spreading of beech in the area. Most likely as a result of human influence via the construction of dirt roads, pastures and settlements resulted in the emergence of areas with open vegetation enabling the easy and direct spread of hazel. .On the other hand if we take into consideration the significant nutrition and protein content of its nut as well as the high level of development of the Bronze Age societies, the ethnographic analogies of hazelnut collecting and the outcome of the palynological analysis of Bronze Age profiles the deliberate human action aiming at the wider spreading of collected plants can not be totally excluded. Since the spreading of hazel seems to be linked to areas lying greater distances from the tells and thus serving as only background areas from the point of view of production during the Middle Bronze Age on the basis of our findings.

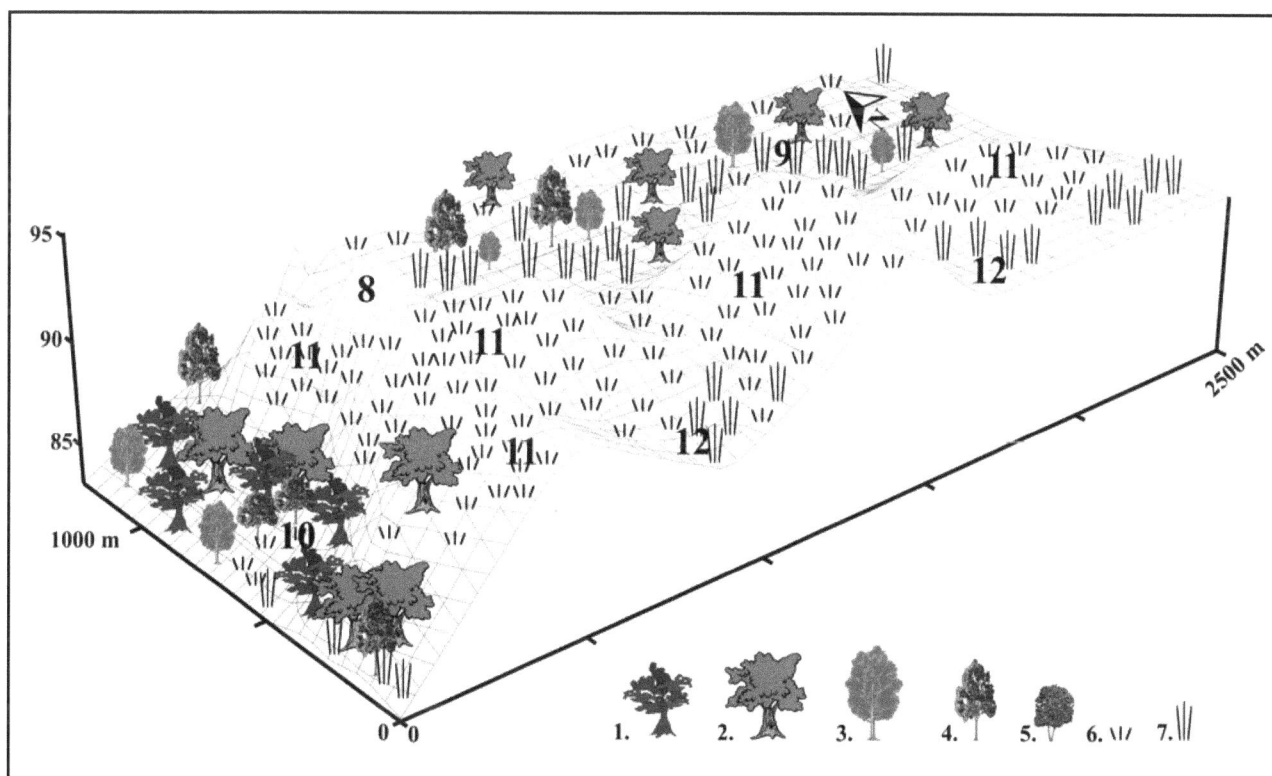

Fig. 3. Reconstructed vegetation around Kenderföld tell at Polgár (Hungary).
1. beech *(Fagus)*, 2. oak *(Quercus)*, 3. poplar *(Populus)* and willow *(Salix)*, 4. hornbeam *(Carpinus)* and alder *(Alnus)*, 5. scrub, 6. grasses *(Gramineae)* within cereals and sedge *(Cyperaceae)*, 7. reed *(Phragmites)* and reed mace *(Typha)*, 8. Kenderföld tell, 9. dry gallery forest with marsh on the buried Pleistocene paleochannel, 10. hardwood gallery forest on the Holocene alluvia, 11. arable land and pastourland on the loess covered Pleistocene levees surfaces, 12. alkalic marshes on the Pleistocene backswamps

In a large number of the pollen profiles the appearance and spreading of grape (*Vitis*) could have been identified during the Bronze Age, being especially significant during the middle part of it. According to detailed analysis of the Hungarian pollen profiles the wild grove vine (*Vitis sylvestris*) was present in the hardwood gallery forests as early as the Early Holocene in the Carpathian basin thus forming a part of the original natural vegetation (FACSAR-JEREM, 1985, JÁRAINÉ-KOMLÓDI 1968, JUHÁSZ 2002, JUHÁSZ et al. 2002). Thus the presence of vine pollens in the Bronze Age part of the profiles is not surprising, though their increased rate might refer to an intensification of their consumption and wine production during the Middle Bronze Age. Although the presence of domesticated noble vine (*Vitis vinifera*) could have been identified even in the Iron Age only on the basis of up to date archeobotanical investigations, we have knowledge of traces and finds referring to wine production from as early as the closure of the Copper Age in the Carpathian basin (GYULAI, 2000). On the other hand finds of vintages, fermentation tanks indispensable for wine production and pruning knives found in Bronze Age localities in the Carpathian Basin. (FÜZES-SÁGI, 1968, FÜZES, 1971) seems to refer to the presence of vine growing even during the Bronze Age (GYULAI, 2000). In other words the increased number of vine pollens and grapes might have been the result of more intensive utilization and deliberate spreading of common wild vine and or the emergence of vine growing. There is no significant increase in the rate of grape pollens towards the tells, however a significant increase can be observed in the number of these pollens in the southern parts of the basin enjoying the influence of sub-Mediterranean climatic effects favoring vine growing and the growth of wild vine during the Middle Bronze Age (JUHÁSZ et al 2002).Thus on the basis of these data, if vine growing was present even in the Bronze Age it was not restricted to or linked to the areas of the tell settlements but the areas with the most beneficial natural endowments for the growing of grape.

The Middle Bronze Age also witnessed the appearance and spreading of the pollen of walnut (*Juglans*). Although its presence can be observed in almost all of the pollen profiles it is mainly connected to the areas of floodplains and the surroundings of the tells. According to the pollen data gained up to date the spreading of walnut growing must have been linked to those Bronze Age groups, which migrated into the Carpathian basin from the Balkan peninsula experiencing its culmination in relation to the Middle Bronze Age tell cultures as it is indicated by the radiocarbon dates.

Besides walnut there has been an increase in the number of grain pollens among the plants grown towards the areas of certain tells. The number of grain pollens was especially significant in the neighborhood of those tells bearing a loessy base rock soil. Besides grain it were the different types of weeds growing along roads, arables and pastures such as pigweeds (*Polygonum*), pearlweed (*Sagina*), tartar orach (*Atriplex tatarica*) prince's feather (*Amaranthus*), rib-grass (*Plantago lanceolata*), plantain (*Plantago major/media*) whose dominance maximum developed during the Middle Bronze Age. These also indicate the emergence of extensive farming and arables as well as stock farming with the construction of developed dirt road network for wagon traffic during the Middle Bronze Age in the area under examination. On the basis of the high rate and increase of grain and weed pollens the tells must have acted as the centers of farming, extensive stock farming and the constructed road network. The rate of the pollens of goose-foots (*Chenopodiaceae*) is outstanding in almost all the profiles of Middle Bronze Age, the most prominent values are those in the neighborhood of the tells showing good agreement with earlier archeobotanical data from the areas of the tells according to which the white goose-foot (*Chenopodium album*) might have served as a potential food source for the Middle Bronze Age cultures.

According to the detailed palynological analysis of sediments deposited in the ditches around the tells, the original vegetation in the neighborhood of the fortified tells has been totally altered. The trees have been fully logged possibly because of defensive purposes. Thus our pollen analytical results seem to reinforce the picture drawn on the basis of previous archeobotanical data, namely that the Bronze Age tells acted as the centers of not only political and social life or trade but also those of agricultural production (farming and stock farming) as well.

SUMMARY

On the basis of the palynological analysis of Bronze Age pollen profiles, mainly those of Middle Bronze Age the presence of highly intensive productive economies could have been identified in the area of the Carpathian basin overwhelming any previous human impacts. On the basis of the intense spreading of grown plants (grains, walnut) as well as those spreading in connection with human activity (wild grove vine and hazel) and weeds an increasing rate of extensive stock farming extending into the areas of alluvial plains and gallery forests may be assumed. This must have been accompanied by increasing farming, corn production and organized collecting (hazelnut, grapes). In relation to the increasing human activities extensive agricultural production must have developed with farming and stock farming, adjacent to the tells in the area and during the period under examination. Thus the Bronze Age tells must have acted as the centers of not only political and social life or trade but also those of agricultural production (farming and stock farming) as well. On the basis of data gained from the analysis of 26 pollen profiles neither the military role of the tells, nor the intensive trade and artisanship could overwhelm the increasing agricultural production of the Bronze Age tells in the Carpathian basin.

Authors' addresses

Pál SÜMEGI
University of Szeged, Department of Geology and Paleontology, 6701 Szeged P.O. Box: 658
Archeological Institute of Hungarian Academy of Sciences, 1014 Budapest Úri u. 49

Imola JUHÁSZ
Archeological Institute of Hungarian Academy of Sciences,
1014 Budapest Úri u. 49

Elvira. BODOR
Hungarian Geological Institute 1043 Budapest Stefánia u. 14.

Sándor GULYÁS
University of Szeged, Department of Geology and Paleontology, 6701 Szeged P.O. Box: 658

Bibliography

BEHRE, K.E. 1988. Some reflections on the anthropogenic indicators and the record of prehistoric occupation phases in pollen diagrams from the Near East. 633-672. In: Huntley, B.- Webb, T. III. eds. *Vegetation History*. Kluwer Academic Publisher, Dordrecht.

EDWARS, K.J.1991. Models of mid-Holocene forest farming in northwest Europe. 133-144. In.: Chambers, F.M. ed. *Climate change and human impact on the landscape*. Chapman and Hall Press, London.

FÜZES M. 1971. Archaebotanical comments on paper of Wine and grapes by Moór E. *Veszprém Megyei Múzeumi Közlemények*, 10. pp. 115-126. (in Hungarian)

FÜZES, M. & SÁGI, K. 1968. The Pannonian roots of vineculture at Balaton. *Filológiai Közlöny*, 14. pp. 347-363. (in Hungarian)

GARDNER, A.R. 1999. The ecology of Neolithic environmental impacts – re-evaluation of existing theory using case studies from Hungary. *Dokumenta Prehistorica*, 26. pp. 163-183.

GYULAI F. 1993. *Environment and Agriculture in Bronze Age of Hungary*. Archaeolingua Press, Budapest.

GYULAI, F. 2000. *Archaeobotany*. Jószöveg könyvek, Debrecen. (in Hungarian)

JÁRAINÉ-KOMLÓDI, M. 1968. The late Glacial and Holocene flora of the Great Hungarian Plain. *Annales Universitatis Scientiarum Budapestiensis - Sectio Biologica*, 9-10. pp. 199-225.

JUHÁSZ I. E., 2002. *A Délnyugat Dunántúl negyedkori vegetációtörténetének palinológiai rekonstrukciója. (Reconstitution palynologique de la végétation depuis le Tardiglaciaire dans la région de Zala, sud-ouest de la Hongrie)* PhD disszertáció Pécs- Marseille, 2002, p.215.

JUHÁSZ, I., DRESCHER-SCHNEIDER, R., ANDRIEU-PONEL,V. & DE BEAULIEU, J.L. 2002. Anthropogenic Indicators in a Palynological Records from Pölöske, Zala Region, Western Hungary. *Abh. Geol. Bundesanstalt. Band* 78. pp.29-37.

KOVÁCS, T. 1977. *Bronze Age in Hungary*. Corvina Press, Budapest.

MAGYARI, E., SÜMEGI, P., BRAUN, M. & JAKAB, G. 2002. Retarded hydrosere: anthropogenic and climatic signals in a Holocene raised bog profile from the NE Carpathian Basin. *Journal of Ecology* 89. pp. 1019-1032.

RACZKY, P., HERTELENDI, E. & HORVÁTH, F. 1992. Zur Absoluten Datierung der Bronzezeitlichen Tell-Kulturen in Ungarn. 42-47. In: BÓNA, I. ed. *Bronzezeit in Ungarn*. Pytheas, Budapest.

SÜMEGI, P. 1999. Reconstruction of flora, soil and landscape evolution, and human impact on the Bereg Plain from late-glacial up to the present, based on palaeoecological analysis. pp. 173-204. In: Hamar, J. -Sárkány-Kiss, A. eds. *The Upper Tisa Valley*. Tiscia Monograph Series, Szeged.

SÜMEGI, P. & BODOR, E. 2000. Sedimentological, pollen and geoarcheological analysis of core sequence at Tököl. pp. 83 – 96. In: Poroszlai, I. –Vicze, M. eds. *Százhalombatta Archaeological Expedition*. Archeolingua Press, Budapest, p.134.

SÜMEGI, P., HERTELENDI, E., MAGYARI, E. & MOLNÁR, M. 1998. Evolution of the environment in the Carpathian Basin during the last 30.000 BP years and its effects on the ancient habits of the different cultures. pp. 183-197. In: Költö, L.-Bartosiewicz, L. eds. *Archimetrical Research in Hungary*. II. Budapest.

WILLIS, K.J., SÜMEGI, P., BRAUN, M., BENNETT, K.D. & TÓTH, A. 1998. Prehistoric land degradation in Hungary: who, how and why? *Antiquity*, 72. 101-113.

www.ingramcontent.com/pod-product-compliance
Lightning Source LLC
Chambersburg PA
CBHW061007030426

42334CB00033B/3401